TRADITIONAL FOOD PRODUCTION AND RURAL SUSTAINABLE DEVELOPMENT

Ashgate Economic Geography Series

Series Editors:
Michael Taylor, Peter Nijkamp, and Tom Leinbach

Innovative and stimulating, this quality series enlivens the field of economic geography and regional development, providing key volumes for academic use across a variety of disciplines. Exploring a broad range of interrelated topics, the series enhances our understanding of the dynamics of modern economies in developed and developing countries, as well as the dynamics of transition economies. It embraces both cutting edge research monographs and strongly themed edited volumes, thus offering significant added value to the field and to the individual topics addressed.

Other titles in the series:

Upgrading Clusters and Small Enterprises in Developing Countries
Environmental, Labor, Innovation and Social Issues
Edited by Jose Antonio Puppim de Oliveira
ISBN: 978-0-7546-7297-5

The Moving Frontier
The Changing Geography of Production in Labour-Intensive Industries
Edited by Lois Labrianidis
ISBN: 978-0-7546-7448-1

Network Strategies in Europe
Developing the Future for Transport and ICT
Edited by Maria Giaoutzi and Peter Nijkamp
ISBN: 978-0-7546-7330-9

Tourism and Regional Development
New Pathways
Edited by Maria Giaoutzi and Peter Nijkamp
ISBN: 978-0-7546-4746-1

Alternative Currency Movements as a Challenge to Globalisation?
A Case Study of Manchester's Local Currency Networks
Peter North
ISBN: 978-0-7546-4591-7

Traditional Food Production and Rural Sustainable Development

A European Challenge

TERESA DE NORONHA VAZ
University of the Algarve, Portugal

PETER NIJKAMP
Free University, Amsterdam

and

JEAN-LOUIS RASTOIN
Montpellier SupAgro, France

Routledge
Taylor & Francis Group
LONDON AND NEW YORK

First published 2009 by Ashgate Publishing

Published 2016 by Routledge
2 Park Square, Milton Park, Abingdon, Oxon, OX14 4RN
711 Third Avenue, New York, NY 10017, USA

Routledge is an imprint of the Taylor & Francis Group, an informa business

British Library Cataloguing in Publication Data
Traditional food production and rural sustainable
development : a European challenge. - (Ashgate economic
geography series)
1. Food industry and trade - Europe 2. Rural development -
Environmental aspects - Europe 3. Sustainable development -
Europe 4. Small business - Europe
I. Vaz, Teresa de Noronha II. Nijkamp, Peter III. Rastoin,
Jean-Louis
338.4'7664'0094

Library of Congress Cataloging-in-Publication Data
Vaz, Teresa de Noronha.
Traditional food production and rural sustainable development : a European challenge / by Teresa de Noronha Vaz, Peter Nijkamp and Jean Louis Rastoin.
p. cm. -- (Ashgate economic geography series)
Includes index.
ISBN 978-0-7546-7462-7 1. Sustainable development--Europe. 2. Rural development--Europe. 3. Food consumption--Europe. 4. Food industry and trade--Europe. I. Nijkamp, Peter. II. Rastoin, Jean-Louis. III. Title. IV. Series.

HC79.E5V297 2008
338.1094--dc22

2008029304

ISBN 13: 978-0-7546-7462-7 (hbk)

Contents

List of Figures

List of Tables

List of Contributors

Alboiu, Cornelia
Institute of Agricultural Economic, Bucharest, Romania

Alexandri, Cecilia
Institute of Agricultural Economics, Bucharest, Romania

Avermaete, Tessa
Departement Rega, Katholieke Hogeschool Leuven, Belgium

Baycan-Levent, Tüzin
Department of Urban and Regional Planning, Istanbul Technical University, Turkey

Brouwer, Floor
Agricultural Economics Research Institute (LEI), The Hague, the Netherlands

Caro, Irma
Research Centre of Science and Food Technology, University of Hidalgo State, Mexico

Couto, João Almeida
Economics and Management Department, University of Açores, Portugal

Figueira, Ana Cristina
Superior School of Technology, University of Algarve, Portugal

Florax, Raymond
Purdue University, Department of Agricultural Economics, USA

Gatti, Silvia
Department of Statistics, University of Bologna, Italy

Gülümser, Aliye Ahu
Department of Urban and Regional Planning, Istanbul Technical University, Turkey

Hockmann, Heinrich
Institute of Agricultural Development in Central and Eastern Europe, Halle, Germany

Ionel, Iuliana
Institute of Agricultural Economics, Bucharest, Romania

Mateo, Javier
Department of Hygiene and Food Technology, University of León, Spain

Mili, Samir
Institute of Economics and Geography, Spanish Council for Scientific Research, Madrid, Spain

Natário, Maria Manuela
Department of Social and Human Sciences, Guarda Polytechnic Institute, Portugal

Neto, Paulo Alexandre
Department of Management, University of Évora, Portugal

Nijkamp, Peter
Faculty of Economics and Econometrics, Free University of Amsterdam, the Netherlands

Noronha Vaz, Teresa de
Faculty of Economics, University of Algarve, Faro, Portugal

Oltmer, Katrin
Agricultural Economics Research Institute (LEI), The Hague, the Netherlands

Ramos, Daphne
Faculty of Veterinary Medicine, National University of San Marcos, Peru

Rastoin, Jean-Louis
Montpellier SupAgro, UMR Moisa, France

Szlanyinka, Edina
Svent István University, Hungary.

Thøgersen, John
MAPP Centre, Aarhus School of Business, Denmark

Vandermosten, Gert
Streekproducten Vlaams Brabant, Belgium

Vőneki, Éva
Agricultural Economics Research Institute, Budapest, Hungary.

Zumalacárregui, José M.
Department of Hygiene and Food Technology, University of León, Spain

Editorial Preface

The worldwide interest in sustainable development has prompted a policy and research interest not only in ecologically-benign developments of key sectors such as industry or transportation but also in the management and assessment of new lifestyles such as healthy food consumption and the sustainable use of products. In this context, agriculture plays a central role as a sector that has a dual character, viz. a high-tech sector producing modern mass products and a traditional sector producing environmentally-friendly goods. Therefore, the agricultural sector is a prominent sector in the current debate on the socially- and environmentally-sustainable development of our planet.

The present volume brings together a collection of new contributions on the interface of agricultural – and sometimes rural – development and the social and economic feasibility of traditional modes of production and consumption in a developed country. It addresses general production issues of the sector, as well as the viability of new consumption patterns or the adoption of traditional agricultural products.

It contains a wealth of arguments on strategies and policies concerning sustainable agriculture, with a particular emphasis on specific goods. The book presents a critical review of the opportunities of traditional production modes in agriculture, and discusses these options from the perspective of both local-regional possibilities and market opportunities, not only nationally but also globally. The general conclusion is that traditional modes of production and consumption in the agricultural sector offer a novel and promising source for a more sustainable development of our societies.

The editors wish to thank many colleagues who have contributed to the success of this collection of articles on traditional food production and sustainable development, not only the authors but also the many referees who carefully judged the quality of the various contributions. The editors wish to thank in particular Patricia Ellman who has carefully checked the manuscript in terms of style and language. We hope that this volume will enhance understanding of the potential of sustainable agricultural production and consumption.

Teresa de Noronha Vaz
Peter Nijkamp
Jean-Louis Rastoin

Introduction

Chapter 1
The Complex Force Field of Traditional Food Systems: Setting the Scene

Teresa de Noronha Vaz and Peter Nijkamp

1.1 Introduction

Europe has many market niches where traditional forms of production are still dominant. The geographic variety of Europe certainly stimulates the existence of diverse niches with a clear local colour and identity. Even in the age of mass production and consumption such pockets of often place-bound products have not disappeared. Rather, they are enjoying a comeback, as the public at large is increasingly looking for goods with indigenous or specific qualities. Traditional production modes – as one of the valuable legacies from the past that are socially and economically active throughout history – appear to have a strong chance of survival. History was responsible for generating several – currently – low-tech activities whose survivors still have a place, albeit sometimes modest, in our contemporary society. This holds in particular for the food sector, which has gained much popularity in recent years. In this book a new vision of the traditional food sector is suggested: it is both a container of valuable knowledge and information and a transmitter of creative production processes over time, based on an established structure of small firms (SMEs), most of them embedded in local environments. It is undoubtedly a challenging task to bring technical and organizational innovation to places where the industrial model can only be applied with difficulty. The attempt to apply it generates debates on local social cohesion and rural sustainability.

The present volume addresses the importance of traditional food production against the background of dramatic changes in the European scene: the urban-rural dichotomy with cities playing a major role as knowledge creators and rural areas fighting for the survival of local identities or for keeping their role as low-tech production areas or as promoters of self-employment in a high-tech low-tech dilemma.

These arguments for the significance of this book are currently receiving ongoing attention, particularly now that the Common Agricultural Policy (CAP) is once more becoming the subject of heated public debate.

Recently, the European Commission has presented its ideas about the medium-term outlook for the CAP and the prospects for rural development in the European Union. The Commission will soon present the CAP Health Check, a package of instruments to transfer funds, before 2013, from agricultural direct payments to the rural development budget. This is a clear sign of the priority given to rural development and the creation of a diversified context for the rural world.

For the moment, the enthusiasm that brought those responsible for agricultural and rural development in the European Commission to confirm that it will be necessary to concentrate more money on rural development policy and refocus CAP is being underlined and justified by new determinants in the world markets: 1) Variations in market prices for cereals have stopped being the effect of surplus crops and instead result from future Indian and Chinese consumer trends; 2) Bio-fuel or other bio-energy programmes can solve part of our energy problems by offering a complementary solution to a global package of alternative energy usage; 3) Expected climate changes, like drought in many regions around the world, and also in Europe, demand fast and reactive market responses to sudden productive catastrophes – the storage and nucleus of local production can represent instruments for covering sudden market imbalances; and, finally; 4) The agriculture of the 21st century needs to be able to manage scarce water resources, not only to show more respect for nature, but also to learn once again to locate production in its natural environment.

Policy makers are using their power to persuade actors to adopt a more responsible attitude and are planning to create, what they call, 'tailor-made' support measures to reduce existing production bias. It is a long-term gamble which requires the proactive involvement of all relevant parties. Single payment entitlements, decoupling, modulation, and cross-compliance are the new instruments for a two-step model in which the European Commission expects to create a grubbing-up scheme as well as to annul planting rights. In this particular context, what is the future for traditional food production?

Traditional food production can have a bright future, providing it is possible to include it in such fixed goals and adapt the sector to benefit from the proposed list of instruments. The wine sector, for example, will have to submit to the end of the planting rights in 2010, and the outlets provided by distillation schemes, which currently allow surplus production will be over. While this represents a gain for good producers, it is a clear loss for the others; on the contrary, pig production will be encouraged to use private storage – a hope for many traditional meat processors, who would not survive if they were not able to maintain very strict quality controls; also, milk quotas are expected to end in 2015 – in this case, increases in dairy production will tend to be directed to exports – a probable advantage for big companies but a challenge for small firms which will need to label and increase the spectrum of new products.

Such changes demand urgent discussions with all the actors involved, including traditional producers, and concerning the respective traditional productive forms. Segmentation, labelling, and other marketing techniques demand more than ever organizational and technical innovation. These innovations can not be neglected by the sector and should be rapidly brought to the fore – not from a central stage but rather from decentralized forums. Our rule as researchers and technicians is to provide analytical tools to better understand the ongoing constraints in this sector and those specific opportunities that in a changing context always arise.

This book centres on the question: What difficulties does the traditional food sector face as it tries to adjust to the world's global trends? Are there any new market opportunities for European traditional food products in the international scene? And if so, how can a rural production region find a clear profile in Europe's sustainable development? In order to respond to these and related questions, a set of scientific articles were prepared by experts from various countries to introduce the reader to the varied world of traditional food production. Most of them address market segmentation and consumption niches as firms' strategic alternatives for survival. In the particular case of food assets, and in spite of the existence of new consumption patterns, consumers' preferences are still moving towards quality standards related to their past memories. Thus, historical identity and cultural heritage appears to have remained as the guardian of such production systems in many regions in Europe.

Most firms that produce traditional foods belong to the low-tech sectors. They perform their business under very difficult global constraints and harsh competitive requirements. So, also in their case, innovation is crucial, and it may promote local prosperity. But this simple observation may become rather complex: if these firms are located in learning regions, they have to meet the prerequisites for technological apprenticeship not only in their historico-cultural past but also in an adjusted and modern business environment. Collective action and network interdependencies can be introduced in order to perform functions such as research, selection, codification, transformation, control and other procedures which together represent very constructive knowledge flows able to transfer know-how to firms and local consumers. Traditional food production may, therefore, account for a significant share of income generating capacity and, most of all, an opportunity for many regions, some of which lagging behind within the European Union or other parts of the world.

1.2 Aims and contents

The book explores the potential importance of the traditional food sector, including its involvement with so many varied consumers across the world. To stimulate this sector to adopt new strategies focussing on innovation and efficiency involves a simultaneous consideration of technological change, social efficiency and environmental responsibility.

This volume is divided into three main parts which map out the most important issues and challenges.

After the introductory Chapter 1 by Teresa de Noronha Vaz and Peter Nijkamp, describing summary the complex force field of traditional food systems, Part I *Sustainability and European Rurality* then proceeds to focus on the justification for a sustainable European rural world, against the background of globalization processes.

Part II *Traditional Markets and Globalization* illustrates how traditional food production has in the past adapted to new circumstances and is now ready to adapt in the future to new consumer requirements and to large scale markets.

Finally, Part III *Mass and Segmentation in Traditional Food Markets* argues that, as a result of marketing restrictions, several food products are introducing specific characteristics that are similar to the traditional attributes, from the perspective of market segmentation, in order to improve product quality and increase sales. In such a strategy, firms acquire local identities and may develop new, sometimes even significant, responsibilities towards the development of their local environment, in line with environmental sustainability.

In conclusion, the book aims to clarify serious questions related to survival strategies and the necessary compromises faced by traditional food production. On the one hand, market pressures and technological innovation call for the rapid modernization of established production structures, while, on the other hand, serious constraints related to social and environmental sustainability require careful reflection and the responsible participation of the firms. The various case studies in this book serve to highlight all the above-mentioned issues.

In the first chapter of Part I, Chapter 2, Jean-Louis Rastoin, in his contribution on '*Is the World Food System Compatible with Sustainable Development?*' critically reviews the intensive agro-industrial model (in particular, the way it is financed, concentrated, specialized and globalized). Although providing remarkable results in terms of product prices and food safety, such a model generates many negative externalities that in the long run will threaten the food equilibrium of the population and the ecological balance of our planet. The author uses the concept of sustainable development to offer a new orientation on which researchers are invited to work: a definition of a new food basis; the design of shorter and more diversified productive and marketing systems; and a proposal for proper governance models on a regional, national and international scale.

As a complement to the previous reflection, next in Chapter 3, '*Sustainability and Agri-Environmental Policy in the European Union: A Meta-Analytic Investigation*', Katrin Oltmer, Peter Nijkamp, Raymond Florax and Floor Brouwer introduce several environmental aspects of agricultural land use, which are closely connected with the justification for the survival of traditional food production. Both the assessment of the environmental effectiveness of agri-environmental policies in the European Union and the need to draw lessons from comparative case-study research in this field constitute the main research tasks of their study, in which meta-analysis is used as a suitable tool for the policy assessment of agri-environmental initiatives in the EU.

Then, within the scope of this first part on rural sustainability, several additional concepts related to sustainability in an EU context are introduced: rurality, governance, and cultural heritage. Rurality is largely discussed in Chapter 4, entitled '*A Comparative Analysis of Rurality at the EU level and Turkey*', where Aliye Ahu Gülümser, Tüzin Baycan-Levent and Peter Nijkamp present a study that aims to compare and evaluate the degree of rurality of the EU Member States, while particularly identifying the place of Turkey, which as yet is not a Member State. Several selected rural indicators are used for comparison and evaluation of 26 countries (EU-25 and Turkey), based on Eurostat and World Bank data.

Multidimensional classification technique and factor analyses are also used to define Turkey's rurality in the European context.

In the Chapter 5, on '*Governance and the Determinants of Local Economic Development*', the authors, Paulo Alexandre Neto, João Almeida Couto and Maria Manuela Natário, argue that the dynamics of territorial governance structures create a virtual geographic space and promote synergies and competitiveness, and therefore the ability to decide, adjust and regulate the agricultural territory depends on competent and efficient governance systems. These systems have an important role to play in the territorial innovation process, all the way down to the choice of projects, regional policies, regulation and/or organization of local activities. The authors present the results of an empirical research project by using an extensive set of firms located in the 'Raia Central Ibérica' Region, in the border area of Central Portugal/Spain, in order to measure the entrepreneurs' satisfaction with the sub-regions' governance systems and their respective impacts on local innovation levels. Their empirical study concludes that the imbalances may be attributed to the lack of common identity, government/public intervention, and cooperation.

The first part of this book ends with a discussion regarding cultural heritage, in Chapter 6. Presented by Edina Szlanyinka, in '*The Role of Cultural Values in Rural Development*', the concept is developed by focusing on the links between rural development and gastronomic tourism. It addresses the economic potential that gastronomy can offer in the development of rural areas, as well as in the possibility to maintain the important heritage brought by local gastronomy. The use of local cultural and human resources and traditional material factors are mobilized in this chapter to defend the role of gastronomy as an instrument of cultural economics and rural development.

Part II of the book addresses traditional markets and globalization. It aims to illustrate how traditional food production has been able to historically adapt, both to meeting consumers' requirements and to entering large-scale markets. The various chapters point out the far-reaching changes that in the traditional food system have undergrown, while at the same time introducing some of its production systems into the industrial-scale model.

Starting with an overview of the old production forms, Chapter 7, entitled '*Meat Processing in Ibero-American Countries: A Historical View*' and written by Javier Mateo, Irma Caro, Ana Cristina Figueira, Daphne Ramos and José M. Zumalacárregui maps out the profound changes in the processing paths in the meat sector that have occurred since pre-historic times. The authors argue that the ancient techniques for preserving meat have been kept enshrined in the cultural food habits of the population, in spite of some changes in the appearance, flavour and other attributes that may have altered traditional products over the centuries. They also speculate on how meat consumption must have been a milestone in human evolution. Drying, smoking, fermentation and salting were prehistoric techniques for preserving meat. Nowadays, in most European countries, a large part of the basis for traditional meat-product processing comes from these cultures, although the appearance, flavour and other attributes of products may have been subject to changes and diversification throughout time.

The next four chapters of Part II illustrate the expansion of internal European markets with respect to large-scale production of traditional food products. The introduction of the euro, intended to foster exchanges within the European Union, also promoted more intense competition between firms through market integration. This was assumed to advance, simultaneously, the specialization and efficiency of European agricultural production. Whether this potential could be exploited depends strongly on the characteristics of the markets, i.e. on the extent of market power or the existence of barriers to external trade. An indication of the degree of EU market integration is given by, for example, the joint movements of the national product prices. Time-series models, especially co-integration techniques, are typically applied for analysing market integration.

As an example of the determinants influencing the supply side of integrated markets, the product of butter is investigated in Chapter 8 '*Market Integration and Market Power in the Internal EU Market for Butter*'. Here, Heinrich Hockmann and Éva Vőneki apply these techniques to the intra-EU trade in butter. They deal with issues such as the extent of market integration in the EU-15; the level of price transference amongst the EU countries and the causes for this phenomenon; the way market integration performs in peripheral regions; and particularly the experiences of the EU-10 during the accession process. The results of this empirical analysis are used for a first assessment of the organization of the internal market for butter. The conclusion discusses to what extent the findings can also be transferred to other agricultural traditional products.

Some European consumption patterns are observed in the subsequent three chapters. Chapter 9, by Cecilia Alexandri and Cornelia Alboiu, describes '*The Romanian Food Consumption Model in the context of the European Union Integration*'. In this case, the peculiarities of the food demand, common to many European less-developed countries, are the central concern. In the description of the Romanian situation, the existence of a dual food consumption pattern is revealed. Food consumption in the rural areas has peasant household production as its main source and is characterized by less diversification, seasonal cycling, and is still based on a subsistence economy. Because of this rural food consumption seems less correlated with the usual economic variables viz. incomes and food prices. The authors point out an apparent paradox in this dual model. Although rural incomes are lower than the urban incomes, food consumption in rural areas is higher than food consumption in the urban areas.

Chapter 10, '*How Could Traditional Consumption Stimulate the Bakery Industry?*', by Iuliana Ionel, offers a complement to the previous discussion and a thorough analysis of one of the leading sectors within traditional production and urban consumption in Romania: the milling and bakery industry. With global investments in technologies surpassing $100 million, firms are now developing new production and marketing strategies. The full segment consolidation process is forcing big operators to expand, either by diversification, or by approaching new market segments. The branch comprises about 6,500 enterprises, of which only 75 produce on a large scale.

Finally, the second part concludes with Chapter 11 on '*Consumer Decision-Making with Regard to Organic Food Products*', by John Thøgersen. The work proposes a method to explore the consumers' level of acceptance of new types of food products in the design of future global markets. A survey in eight European countries is used to present a model of consumer decision making and behaviour with regard to organic food. The author found that the justifications given and the reasoning behind choosing organic products are quite similar across countries, but that behavioural intentions are predictive of behaviour only in the northern European countries; this not being the case in southern Europe. This is an interesting framework to better understand the difference in consumption attitudes between northern and southern Europe.

Part III of this book is also very challenging, because it is based upon the evidence that, as a result of marketing restrictions, many production modes, among which several food products, are based on specific characteristics that are similar to traditional attributes in a segmentation strategy geared to better sell or improve product quality. In this process, firms acquire local identities and may develop new, sometimes even strong, attitudes towards growth and development in a local setting.

To confirm these general arguments, in Chapter 12, Teresa de Noronha Vaz analyses '*Local Honey Production: Export or Indigenous Growth?*' She offers a brief overview of the situation related to the European production and trade of honey. This essay draws attention to the limited opportunities related to the specific market for honey production and trade. Even though within the EU all Member States have to import honey to meet internal consumption, this product is subjected to different trends in its production process. A significant amount of honey is produced under conditions of scale economies, but much of the growing demand is directed to very narrow market segments, characterized by specialization requirements and high quality standards. Honey serves as an excellent example of technical innovation in a traditional food product system, demonstrating the importance of sales guarantees in commercial circuits and long-term contracts with small producers.

There is a subtle continuity between Chapters 12 and 13: honey production is a less regulated market and has a more disperse production process than the contemporary olive oil production process. In '*Market Dynamics and Policy Reforms in the Olive Oil Sector: A European Perspective*', Samir Mili explains the degree of maturity of the olive sector in Europe and presents an overall evaluation of the market change and policy reform processes. Using a SWOT analysis, the study investigates the economic profile of the sector and its expected evolution. The analysis is performed with a systemic approach to market and policy changes, under the general hypothesis that they are governed by a series of economic and regulatory factors for which little empirical evidence exists, or which still are the subject of a public debate whose outcomes are not clear. It is also assumed that, while some of these factors act at a global level, others may vary across countries and even within countries amongst different players, which means that there is a wide spectrum of possible strategies and courses of action for the future of many producers and regions, located in the Southern part of Europe, which rely on the incomes originating in this sector.

There are a number of conflicts emerging from the coexistence between scale production and market segmentation. The pressure of very large companies and multinationals drives small firms to exploit the historical and cultural values of products that are embedded within a specific region. This phenomenon is very explicitly pronounced in the Belgian brewery sector, where small speciality brewers have to find strategies to survive in the shadow of the world's largest brewer. Chapter 14 explores the strengths and weaknesses of the regional initiatives for European labelling as tools to safeguard traditional methods. In '*Traditional Beers in a Global Market Economy*', Tessa Avermaete and Gert Vandermosten describe the struggle of the small firms fighting for advantages in a competitive world. The study is based on general facts and figures of the brewery sector and in-depth interviews with brewers and experts.

From the perspective of industrial modelling, there is a dynamic progress that is particularly expressed in Chapter 15 and is related to the different levels of maturity in the food production processes. To illustrate this aspect, the chapter on '*Protected Designation of Origin, Sustainable Development and International Policies: A Survey of DOC Wines from Emilia-Romagna*', by Silvia Gatti, revisits the wine sector as a case of international competition facing the recent breakthrough out of the 'New World'. The very detailed overview of the multiple actions taken in Emilia-Romagna to guarantee environmental sustainability in the region during the process of wine certification is based on available data related to organic farms and enterprises. The wine, produced in one of the most well-developed traditional production processes and a fully maturated traditional product in international markets, is analysed as an environmentally-benign product, active in promoting regional development.

In Chapter 16, to close the third part of the book, Teresa de Noronha Vaz and Peter Nijkamp, provide some evidence to justify the renewed social role of segmented markets. In '*Large-Scale Production and Market Segmentation: An Uneasy Relationship*', theoretical and empirical references are used to justify emerging organized diversified structures that may comprise community supported agriculture, farmers' markets, urban agriculture, and regional food labels.

Finally, serving as a general conclusion to the book, Chapter 17 '*Traditional Food Production, Market Segments and Rural Sustainable Development: A Synthesis*', by Jean-Louis Rastoin sheds light on some of the doubts related to the necessary compromises faced by traditional food production: if, on the one hand, market pressures and technological innovation call for modernity in the productive processes, on the other hand, the constraints related to social and environmental sustainability in Europe require new approaches and creative interfaces.

PART I
Sustainability and European Rurality

Chapter 2

Is the World Food System Compatible with Sustainable Development?

Jean-Louis Rastoin

Introduction

The food system is '*the manner in which Mankind organizes itself in time and space to obtain and consume its food*' (Malassis, 1994). This food system has passed through different stages in the history of human societies and countries. It has always (for 10,000 years or a few centuries, depending on the country) been founded on agriculture with the development of farms, based on blood ties and an attachment to Mother Earth, which act as the sites of both food production and consumption. There is unity of location, an autarchy, an ultra-short commodity chain. This stage of development is still widely present today, in the poorest countries. Currently, it concerns several hundred million people throughout the world.

This stage lasted until the 18th or 19th century and was followed by the division of labour (the farmer, the craftsman, the trader) and urbanization, which have broken down the food chain (from the field to the plate or glass). The food system then forms an interactive umbrella for the production of inputs, the production of agricultural raw materials, the transformation of these raw materials into consumable products and their marketing, as well as for all the services necessary to support the different channels (transport, finance, research, training and administration). This multiplicity of activities and actors and the importance of the function of food mean that, in most countries, the food system has become the leading economic sector by number of jobs and turnover.

About half a century ago, the food system entered the agro-industrial age with a generalization of the industrial method of production (i.e. primarily standardization and large-scale production) and mass consumption. This agro-industrial stage is characterized by an extraordinary extension of the agri-business channel and a major reduction in the time devoted to preparing and eating meals.

Since the end of the last decade, we have begun to see the emergence of a 4th food age which we refer to as 'agro-tertiary', as the foodstuffs tend – from the point of view of their economic content – to become services rather than material goods. Thus, in the United States, almost half of the final price of the average food product comprises services or payments: transport, marketing (publicity represents more than 10 per cent), banking interest and insurance, distribution margins, taxes and profits. The part corresponding to the agricultural raw materials has fallen to 10 per cent. The rest, totalling 35 per cent, is paid primarily to the agri-business industry and the packaging companies. As for consumption, half of the average household budget devoted to food is spent in restaurants, largely dominated by fast-food outlets.

The agro-industrial model is experiencing rapid growth in the emerging countries (with intermediate income), stimulated by the expansion of large volume distribution: in Latin America and South-East Asia, supermarkets now control 50 per cent of retail trade compared with 20 per cent ten years ago. Indeed, in the food-processing channels, the downstream concentration causes the standardization of products in accordance with the distributors' norms and a rapid restructuring of the agri-business industry and the agricultural sectors upstream.

Whatever the configuration of the food system, food remains the basis not only for life but for the social act which is (or was) the sharing of a meal (Fischler, 1990) and, to a large extent, for society in general, as all the great civilizations, be they Mesopotamian, Egyptian, Chinese, Mayan, etc., have a strong bond with agriculture. In terms of human development and social structure, the objective is therefore fundamental.

To answer the question asked in the title of this chapter, we will first describe these characteristics of the food system, which would tend to become generalized worldwide in a 'go with the flow' scenario. Then we will consider the possibility of an alternative scenario, insofar as the dominant model leads to a dead end from the point of view of sustainable development. In conclusion, we will examine a number of ideas, which might facilitate the transition towards a 'desirable model'.

2.1 Characteristics of the dominant model: The agro-industrial and agro-tertiary food system

We will consider the food system from the point of view of consumption and of the production-marketing mechanism. From the point of view of consumption, the agro-industrial model, despite its undeniable contributions to which I will return later, has not succeeded in achieving the objective of any food system on a global scale, as defined by the World Food Summit organized under the auspices of the FAO in Quebec in 1995: '*To ensure universal access to food available close at hand which is economically accessible, culturally acceptable and satisfactory in health and nutritional terms.*'

With regard to the final point, more than 50 per cent of the world's populations, representing 3 billion people, were suffering one or other form of malnutrition at the start of the 21st century, according to WHO studies, with women and children the most affected (Delpeuch et al., 2005).

In particular, we should note that, today, 850 million human beings suffer from undernourishment, a phenomenon which is concentrated at a level of 95 per cent in developing countries.[1] The resulting costs are colossal: several hundred billion dollars due to premature deaths, the loss of productivity, absenteeism at school and at work, etc. (FAO, 2005).

At the same time, almost 30 per cent of the population of the United States and 20 per cent in Europe suffer from obesity (BMI, body mass index > 30 kg/m^2). In total, there are more than 300 million overweight people in the world, i.e. over-nourished. This phenomenon also increasingly concerns developing countries. This dietary deviation causes terrible pathologies referred to as non-transmissible or chronic diseases linked to diet, which are the first cause of death (cardio-vascular illness, diabetes, cancers of the digestive tract, osteoporosis) and generate considerable economic costs (€6 billion in France, and at least $90 billion in the United States in 2000).

The causes of this 'dietary disorder' have been identified: poverty; the status of women; deficiencies in the health systems; an absence of education; and the lack of public policies dedicated to the issue of food (Sen, 1981). With regard to this final, very important, point let us recall that the European Union White Book on diet dates from 2000 (and it is far from being a food policy), and that the national nutrition-health programme (PNNS) only started – feebly – in France in 2001.

The production model of the agro-industrial age can be qualified as intensive, specialized, concentrated, financialized and on the road to globalization.

Intensive: as the technical outputs are very high per ha for agriculture, per m^2 of factory or supermarket for the food industry, and per worker in all three cases. For example, one ha of irrigated land can produce nearly 20 t of corn; a cheese factory can produce 150,000 pasteurized *Camembert* cheeses per day. An employee in the oils and fats industry generates an average turnover of more than €800,000.

Specialized: as a result of the selection of a small number of plants and species of animals in contemporary agricultural systems. Scientists estimate that, from a potential of 30 thousand edible plants, barely 120 are commonly grown and only 9 cover 75 per cent of the dietary needs of the world population, of which 3 (wheat, rice and corn) represent 60 per cent (Raoult-Wack, 2001). We are far from making full use of biodiversity! The model is also specialized with regard to the food industry, which today results in an industry which assembles ingredients from the four corners of the world according to relative costs. By adding the distances covered by all of the elements necessary for its production and delivery ('food miles'), we have calculated that a pot of yoghurt contains '8,000 km' of transport. The externalities costs, which can be imputed to the transport of food products to the United Kingdom, have been estimated at 7.4 billion GBP for a total of 30 billion km covered in 2002 (Smith et al., 2005). With the prospect of a major increase in transport costs in the coming years, the limits of such a productive model are clear for all to see.

1 There are 815 million undernourished people in developing countries, 28 million in transitional countries, and 9 million in industrialized countries, for the period 2000–2002.

Concentrated: as in France, for example, 2/3 of agricultural production is manufactured by less than 1/4 of the farmers; 2/3 of the food industry turnover is generated by less than 10 per cent of firms; and 90 per cent of the self-service retail food trade is provided by six companies. The very high concentration of 'super-purchasing outlets' gives huge market power to the multinational groups active in large-volume distribution: IRTS (Auchan and Casino), Agenor (Intermarché, Eroski and Edeka), CMI (Carrefour).[2]

Financialized: because the leading agro-industry and large-volume distribution firms are all listed on the stock exchange and are therefore subject to the will of their shareholders who, more and more, provide funds, and whom the managers treat as investors rather than as industrialists. These managers dictate the rates: short-term growth and profitability. Governance is based on shareholdings, not on partnerships (Pérez, 2003).

On the road to globalization: as we witness a triple phenomenon characterized by the growth of international trade, the development of FDI (foreign direct investments), and the diffusion of the Western model of consumption by the mass media. The international trade on food goods has increased twice as fast over the past half century as production (about 4 per cent per year for world exports of food products compared with 2 per cent for production, according to FAOstat). In 2004, the export/production ratio stood at about 15 per cent with peaks rising to 75 per cent for stimulant needs (coffee, cocoa, tea). Trade is concentrated among certain operators: multinational firms generate 2/3 of international transactions, either because it is inter-firm trade (between subsidiaries belonging to the same group), or because one of the operators is a multinational. FDI increased considerably during the 1990s, in particular in the large volume distribution sector (Carrefour has over 10,000 stores in 50 countries throughout the world) and the agri-business industry sector. Finally, the promotion of products from the agro-industrial TNC on television, thanks to huge advertising budgets,[3] broadens the markets for a limited number of brands and products which tend to become global, thereby contributing to the standardization of the consumption model in accordance with the standards of the TNC, which rarely correspond to those of nutritionists.

The characteristics of the agro-industrial production model mean that this model generates negative externalities, i.e. problems or malfunctions, the costs of which it does not at present cover, and which therefore hold little sway in the strategic decisions of the dominant actors. This might include the exhaustion of natural resources and the degradation of the countryside, the hyper-specialization of production units and the artificial hyper-segmentation of products, which all exacerbate the economic disparities between firms and between consumers. Furthermore, international trade liberalization and the low cost of transport goods

2 These structures are, for the most part, based in Geneva in order to avoid European Union regulations concerning competition.

3 More than $17 billion for the world top 20 firms in the agri-business industry, representing almost 5 per cent of their turnover in 2002 (Ayadi et al., 2005).

are leading to the delocalization of activities towards sites which benefit from advantageous comparative costs and from which the products are exported throughout the world. For example, the standard frozen chicken produced in the United States or Brazil for less than 1 dollar competes with native Moroccan or sub-Saharan African chickens, thereby destroying small local producers, who will then swell the numbers of unemployed in the cities, and distorting the organoleptic typicality of traditional preparations, thereby causing the regional culinary heritage to disappear in the long term.

2.2 Is another model possible?

Having examined the agro-industrial and agro-tertiary model and before considering the feasibility of an alternative model, it is essential that we consider what this model has demonstrated.

First, this model banished the spectre of famines and allowed a global level of self-sufficiency to be achieved: if all the inhabitants of the planet shared world food production equitably, the nutritional standards would be satisfied today. The last famine caused directly by food was suffered in Ireland in the middle of the 19th century, resulting in more than a million deaths. It is true that the 20th century was the most deadly in world history, and that the victims of hunger could be counted in tens of thousands (China, USSR, Africa). However, these famines were primarily political or military in origin and only secondarily the result of natural catastrophes (cataclysms, floods, droughts) (Devereux, 2002). Technical progress has been decisive in this quest for self-sufficiency. In four decades (1961–2002), the average world output of rice has doubled (2 to 4t/ha), while that of wheat has tripled (1 to 3t/ha). This means that the agronomic sciences have enabled us to perfect an operational food production system which is very efficient in relation to the objective of self-sufficiency.

Second, we should attribute a very high fall in food prices to the agro-industrial system (if we take the consumers' point of view): in France in 1700, 300 hours of work were required to buy 100 kg of wheat, while barely 2 hours were sufficient in 2000. This evolution was made possible by the fantastic gains in agricultural productivity and in the agri-business industry. We know that the fall in food prices frees purchasing power for other goods and services and contributes to economic growth.

A third conquest is food safety. Despite recent high-profile crises (mad cow, dioxin, listeria, etc.), we note that the number of deaths due to the toxicity of foodstuffs is now very low: the agro-industrial food system is very safe, even if it has become vulnerable to infectious pathologies due to its concentration.

A fourth positive element is the effect of the agro-industrial system on economic activity. As a result of its sophistication, it has allowed new sectors, such as packaging, logistics, distribution and catering, to be created or to experience a boom. Consequently, employment has been maintained, whereas it has slumped in other sectors. The destruction of agricultural jobs has been accompanied by the creation of posts, primarily in the service industries. With a total of 4 million in France and 16 million in the United States, the number of personnel employed in the food system has suffered only a slight fall in the long run.

Finally, the society of abundance, which characterizes certain countries, provides hyper-choice and therefore hedonistic satisfaction through consumption.

At the end of this analysis of the dead ends and successes of the agro-industrial model, we can consider its durability. This type of question can only be considered globally, as the world is now a village, at least with respect to every form of communication and trade. Before we even tackle the, albeit, essential problem of nature, we must first consider the human element.

The world population is set to increase by a further 50 per cent by 2050, after which date demographers forecast a stationary state at around 9 billion human beings. The answers to the question: Can the Earth feed 9 billion human beings? In simple terms, we can answer 'yes' on a technical level[4] and 'yes if' on a socio-economic and political level. Solutions exist in research laboratories with regard to the agro-industrial model. Nevertheless, the diffusion of these solutions in developing countries would require the implementation of major policy (in particular the allocation of wealth) and organizational reforms, as well as massive investment, unparalleled in relation to the financing currently awarded by rich countries in favour of poor countries (Bruinsma, 2004).

Furthermore, a revision of the consumption model would prove essential. The major trends we have seen at work these past 50 years will become subject to a dietary regime based on a high level of consumption of animal proteins which are very costly to produce (7 plant calories are required to generate one animal calorie). This means that the model cannot be extended to the world population, as it would require land and water in excess of the total resources available on the planet. Nutritionists also condemn this model because it includes the consumption of excessive lipids and sugars (i.e. 'empty calories'[5]) and generates large-scale pathologies; the only other scenario that we could envisage is a change in dietary behaviour.

According to doctors, it is necessary to move towards a more natural and diversified diet. This diet presents the advantages of not only preventing non-transmissible or chronic diseases linked to diet but also of being more agreeable to the senses and of restoring the social and inter-generational link, which is sadly missing in individualist societies.

Although it is possible to provide a precise definition of the consumption model towards which we should tend, nevertheless, the question of the food production model is by no means solved. For many years, professionals and

4 Thanks in particular to the prospects offered by biotechnologies.

5 'Empty calories' means low cost as a result of technical and managerial progress and agricultural policies favouring oleoproteaginous products (soya in the United States) or sugar (European Union Common Agricultural Policy).

associations have warned us of the dangers of the agro-industrial model and recommended an alternative model, based on short commodity chains and firms operating on a human scale. Nevertheless, this approach does not include either economic calculation or the idea of time. Despite the empathy it might encourage in the context of huge agro-industrial and agro-tertiary firms, a productive model based on small agricultural and traditional production units would necessarily lead to a drastic fall in labour productivity (and even the productivity of land and equipment, for technical and economic reasons). We should know that, today, one French farmer feeds almost 80 people, of whom 70 live within the national borders, and an employee in the food industry supplies 125 consumers, of whom 100 are in France. In other words, less than 10 per cent of the active population in rich countries work in the production of foodstuffs. Moreover, in many countries, the food system is highly integrated in the international market, which means that a large, and sometimes vital, income comes from abroad. Consequently, a fall in export capacities due to lower international competitiveness would be detrimental to economic growth and employment.

The model inspired by the slogan '*Small is beautiful*' (Schumacher, 1973) which appeared after the first oil crisis, implies five socio-economic consequences in those countries which have reached the agro-industrial stage:

1. A significant increase in the active population working in the agricultural and agri-business sector (need to find volunteers);
2. A major increase in food prices (need to devote a larger part of the household budget to food);
3. Greater time invested in preparing meals (rather than using 'ready-to-eat' products);
4. A fall in currency resources;
5. The redefinition of the occupation of space by slowing the exodus towards coastal areas and by ensuring a more equitable distribution of material infrastructures and services.

These five consequences or conditions would signal a break from the trends observed for the past century in most countries throughout the world. Other changes would be necessary, which also pose certain problems: rebalancing territories with new methods of marketing products (short commodity chains). Some authors describe this new configuration as an 'alternative model'. The forecasted model would be a combination of various alternative models (Winter, 2003; Watts et al., 2005).

We can try to quantify the potential size of the alternative model from the French case. In France, the trend of the food market indicates segmentation in three categories as shown Table 2.1.

In the future, the food industry TNCs will capture the innovative market of nutraceutics and ingredients such as Omega 3 to maintain their domination on the market, according to consumer preference. After the food safety crisis, the

TNCs tried to use the concept of 'Terroir'[6] as a marketing argument. But there is a contradiction between those very large-scale globalized firms and Terroir products, because this kind of product is contingent with SMEs, territory anchorage, historical and cultural aspects. Because of that, it may be expected that Terroir products will return – sooner or later – to SMEs. In France, this market is significant and it has experienced a high growth rate. It is possible, with an adequate public policy, to increase its market share.

Table 2.1 French food market segmentation and trends, 2004

Segment	Sales (€ Billion)	Market share	Annual average growth rate
Mass Market	97	75%	0–1 %
Innovative Market	6	5%	15–20 %
'Terroir' Products Market	26	20%	5–10 %
Total	129	100%	1–2 %

Source: Own estimates.

In developing countries and in particular the least developed countries, the situation is radically different. Over time, in most countries throughout the world, we have seen the emergence of a dual, if not schizophrenic, food system. On the one hand, a sub-system oriented towards the middle class and the wealthy in the major urban centres, and exports which reflect the agro-industrial model; on the other hand, a traditional sub-system in the agricultural or traditional stage which concerns the majority of the rural areas. In general in these countries, the agricultural population is numerous; relative food prices are high (they absorb a large part of the household income); a considerable amount of women's time is devoted to preparing the meals on account of the low level of development of food products; and integration in international trade remains weak, with the exception of some rare commodities.

In these countries, the priority is naturally to fight poverty through the modernization of agriculture and the diversification of activities. These countries should avoid copying a model whose limits we now see, and instead include sustainable development objectives in their policies. This means, on the one hand, an institutional change in the organization of the people involved in the national food systems and of international cooperation measures; and, on the other hand, specific handling in international negotiations of both the protection of the channels and market access.

At the end of this diagnosis, we can state that the world food system (be it agricultural, traditional or agro-industrial and agro-tertiary) does not satisfy the

6 'Terroir' is the territorial identity of a food product.

recommendations required for sustainable development. If, at a global level, it succeeds in providing low-price commodities (economic efficiency), it is often to the detriment of the natural environment (negative externalities) and social justice. It does the latter by generating social injustices between countries and, within each country, between the actors in the commodity chains, be they producers, traders or consumers.

2.3 Which public policy?

The food system is an excellent illustration of the hybrid governance highlighted by Williamson and the neo-institutionalists. This can initially be explained by the nature of the good produced and consumed: food is both a source of life and full of psychological and socio-cultural attributes. That is why it is closely monitored by the numerous entities active in the food system: first, the public authorities, as food can cause health problems; second, professional corporations, as a deviation from the codes of good conduct within the trade may prove dangerous to the profession as a whole; third, the consumers who desire reassurance with regard to the food they eat; fourth, the employees' unions, as economies of scale – and therefore restructuring – continue to present a threat to jobs; and, finally, local communities as, in rural areas, agriculture and the food-processing industry often represent the last bastions of economic activity. Numerous institutions 'supervise' the firms involved in the food system: global regulatory bodies, which are still somewhat modest (*Codex Alimentarius* FAO-WHO, WTO); a plethora of community regulations (product definitions, quality standards, provisions relating to competition); national legislation, which is no less abundant (in France, the law concerning the 'repression of food frauds' dates back to 1851); and a multitude of private conventions and contracts governing the relations between operators. However, does this hybrid governance satisfy the new demands of sustainable development?

Coined in 1990 by Gro Harlem Brundtland, Prime Minister of Norway and President of the World Commission on the Environment and Development, the concept of 'sustainable development' was made official at the international conference in Johannesburg in 2002. Long confined to the small circle of militant ecologists, it has since been largely mediatized and taken up by firms and governments alike (Boutaud, 2002). Primarily concerned is the food industry, as it takes its products from nature and provides its products to people. Having demonstrated the limits of the agro-industry model and suggested the outlines of an alternative model, we will now attempt to sketch the outlines of a public policy of 'sustainable food development' (SFD) based on three objectives:

- A balanced diet for all (equity);
- A production model which respects the environment (ecology);
- 'Socially responsible' economic efficiency (economy).

2.3.1 Providing consumers with a balanced diet

Since the pioneering work of Cépède and Lengellé and Josué de Castro and Sen, we know that the main cause of malnutrition is poverty. Strategies to tackle undernourishment first and foremost require the reduction of poverty and as a result primarily concern the rural population, representing 2.5 billion people in developing countries who earned their livelihood from agriculture in 2000. The task is enormous, as the modernization needs of agriculture in terms of human, technical and financial resources are quite considerable. Over the past 10 years, however, we have observed a stagnation of PDA (public development aid granted by the OECD countries), without the gap being filled by FDI (foreign direct investment, coming mainly from multinational firms), as these firms do not view agriculture as a worthwhile investment, both for political reasons and in terms of profitability. In this field, then, there is no behaviour within the international community, which corresponds to a policy of 'SFD'. In this case, we must recommend a multilateral public policy in favour of mobilizing funds, knowledge and skills in favour of the least developed countries with a view to reducing poverty. This is the recommendation of the 'Millennium Assessment', launched under the auspices of the United Nations.

At the opposite end of the scale, excess food consumption causes health problems leading to illnesses, malaise and high costs to society. About 80 per cent of the effects of over-nutrition can be seen in rich countries and 20 per cent in developing countries. This calls into question the hyper-calorific and hyper-carbohydrate diets which can be observed worldwide due to their being promoted by multinational companies and a highly mediatized cultural model. Once again, 'socially responsible' measures prove indispensable, be it with regard to individuals, the family, schools or firms. Private operators will not take these measures spontaneously. Public initiatives must be introduced, such as the National Programme of Nutrition and Health (NPNH) launched by the French government, but with the allocation of significant financial resources. This type of programme must first consist of informative (via the mass media) and educative (in schools) initiatives, followed by incentives (in the form of regulatory recommendations) for the food industry to reduce the harmful nature of certain food components (sugars and fats) and finally direct aid to the consumer categories which are financially disadvantaged (for example, in the form of tokens for selected foods).

2.3.2 Producing while respecting the environment

We mentioned earlier the damage caused by agricultural and zootechnical intensification: destruction of 'carbon wells' through deforestation; erosion of arable land; pollution of foodstuffs, soils and groundwater by synthetic chemical agents; reduction of biodiversity by the selection of plant and animal varieties; high-density, industrial breeding generating harmful effluents and stress amongst animals; deterioration of the visual landscape; desertification through the concentration of farms and food-processing plants, etc.

Furthermore, agriculture is the leading sector in terms of water consumption, and we are heading towards a water shortage on a global scale. Finally, the exhaustion of fossil fuels raises the question of prioritization of their use (the provision of food is certainly a priority) and replacement (by biomass fuels which require large areas of agricultural land could pose a threat to food crops). Public policies aimed at protecting the environment must therefore take into account three different elements – i.e. land, water, and chemical inputs – throughout the food chain by defining priorities for using resources which are becoming scarce (water and land), supporting the emergence of new production models, and introducing the traceability of foodstuffs.

2.3.3 Which economic efficiency?

The economic efficiency of the food system must be considered from the dual point of view of production and consumption, on the one hand, and general and individual interests on the other.

For firms, profit is the source of investment and determines long-term survival. Generally, if we examine the differential between the total cost of foodstuffs and their sale price, the food system succeeds in generating significant net profits. Net value is therefore created within the food system. However, the distribution of this net value is far from equitable: weak, and even negative, with regard to agriculture, it is more comfortable downstream if we assess it through the results of the major firms. This situation justifies public intervention through taxation in order to redistribute the profits.

When it is deflated or even more so when it is expressed in hours of work, the price of food has continued to fall over a long period. This means that the productivity gains in the food system (in particular in agriculture and the food-processing industry) have been considerable and have largely benefited the consumers. Today, it would seem that we are nearing an asymptote in this field, although consumers' demands, relayed by the supermarkets, are still pressing in terms of food quality/safety or information/communication/convenience. The food-processing industry and agriculture both run the risk of finding themselves in a managerial impasse. There are two means of relaxing the constraint: the supermarkets should relax the pressure they exert on their suppliers, and the consumer should be willing to pay as much for food as for medicine. In both cases, public incentives would seem to be indispensable.

In short, the aim of making the different entities active in the food system aware of their social responsibilities with a view to supporting sustainable development can not be achieved simply by applying the teachings of market theory. Public and professional regulations, individual auto-regulation, new forms of organization, and new models of production and consumption are all necessary. From the point of view of political economics, this proliferation of regulations poses a problem of consistency and coordination (Godard, 2001). Sustainable food development implies a new public policy, making food for people the heart of our concerns, while respecting the demands of social equity and the preservation of our natural heritage.

2.4 Conclusion: Towards a transition model?

Through a critical examination of the agro-industrial model, we arrive at the conclusion that a 'go with the flow' scenario, i.e. the prolongation of past trends in both the field of consumption and that of food production, is not 'sustainable'. We have demonstrated that the 'Western' consumption model is not only damaging at a personal and societal level but also cannot be extended to cover the entire planet. As for the production model, which is highly 'asymmetric and predatory', its negative impact on the physical and social environment, and the fact that it is managed in accordance with purely financial reasoning, means that it too cannot be generalized.

Nevertheless, a return to the 'traditional' age can hardly be envisaged on account of considerations which can be sociological (we live in a society of individualistic mass consumption, the bases of which must evolve), economic (need for low-cost production and international integration) and technical (all our R&D is oriented towards performance based on economies of scale and therefore implies large production units).

It is therefore essential that we consider the means of organizing the transition towards a new model of 'sustainable food development', i.e. in accordance with the three objectives of economic viability, social equity, and ecological viability. Because of the extreme diversity of situations observed, this model can only be hybrid in nature, combining modern (based on globalization) and post-modern (based on territorial attachment) configurations according to geographic areas, mentalities and kinds of behaviour.[7]

This means that we cannot reckon with only regulation by the market. A genuine food policy must be implemented which at present is not visible in any country in the world (Rastoin, 2005b).

A food policy must be an efficient incentive to improve the nutritional regime. It is legitimized by considerations concerning public health (prevention of illness, well-being) as well as economic concerns (lowering of the direct and indirect costs of pathologies). Fundamentally, it must be based on a modification of the consumer's behaviour through education from an early age. It requires reflection on the allocation of budgetary resources (revaluation of food prices) and time (increase in time devoted by households to preparing food and taking meals). It should also guide agricultural and industrial policy with regard to improving the nutritional

7 According to North, the process of economic development depends on four factors: the quantity and quality of human beings; the stock of knowledge; the institutional framework; and the system of beliefs (North, 2005).

quality of the products sold to consumers and remodelling the production-marketing model by means of diversification and shorter commodity chains. Finally, it must include an effort towards R&D vis-à-vis these models, in particular the technical developments, the baskets of products, and the formats of the companies.

As a result of the intensity of trade between countries, such a food policy involves regional coordination and international dialogue (WTO). In light of the deviations caused by the globalization of the agricultural and agri-business markets, we might envisage 'regionalizing globalization' in order to 'relocalize' the food systems. This would involve closing the distances between the production and consumption sites. Restoring short commodity chains would have the advantage of maintaining (if we are not too late) the diversity of consumption models (by causing them to evolve towards improved nutritional suitability[8]), and maintaining or creating activities, and therefore jobs, in rural areas in most countries throughout the world.

The alternative model must not fall into the trap of the old-modern debate which has been running for more than 2000 years, at least with respect to agriculture. As early as the 1st century AD, the 'first' agronomist, Columella, a proponent of intensive and specialized large-scale agricultural cultivation, disagreed with Pliny the Elder who encouraged the diversification of production in the context of the family farm (Padilla et al., 2005). It is indeed necessary to invent a new agri-business model, which reconciles the historical heritage specific to each society with the scientific and technical knowledge of our own century. The food system could, then, be the field to be favoured in order to initiate the essential changes on which the quality of our future depends, as suggested by the project of the philosopher Ernst Bloch (1885–1977) '*...to make the earth and the attachment to the land into contemporary elements, the foundations of freedom and the need for belonging as well as vectors of sensitivity and a universalist conscience*' (Alphandéry et al., 1992). The earth is therefore one of the essential foundations of sustainable development.

References

Alphandéry, P., Bitoun, P. and Dupont, Y. (1992), *L'équivoque écologique*, Paris, La Découverte.

Ayadi, N., Rastoin, J.L. and Tozanli, S. (2005), 'Les opérations de restructuration des firmes agroalimentaires multinationales entre 1997 et 2003, Agrodata', Paris, Agia-Alimentation, 2004 Working Paper, Montpellier, UMR Moisa.

Boutaud, A. (2002), 'Développement durable, quelques vérités embarrassantes', *Economie et Humanisme*, n° 363, décembre.

Bruinsma J. (2004), *World Agriculture, Towards 2015/2030, An FAO Perspective*, London, Earthscan Publications Ltd.

8 See also the excellent analysis by Rémésy (Rémésy, 2005).

Delpeuch, F., Le Bihan, G. and Maire, B. (2005), 'Les malnutritions dans le monde: de la sous-alimentation à l'obésité', in: G. Ghersi (dir.), *Nourrir 9 milliards d'hommes*, Paris, ADPF, pp. 32–37.

Devereux, S. (2002), 'Famine in the Twentieth Century', IDS Working Paper, 105, Brighton, University of Sussex.

FAO (2005a), *The State of the Food Insecurity in the World*, 2004 (SOFI), Roma.

FAO (2005b), *Base de données FAOstat*, Roma.

Fischler, C. (1990), *L'Homnivore*, Paris, Odile Jacob.

Goddard, O. (2001), 'Développement durable: exhorter ou gouverner ?', *Le Débat*, N°116, septembre-octobre, pp. 64–79.

Malassis, L. (1994), *Nourrir les Hommes*, Paris, Dominos-Flammarion.

North, D. (2005), *Understanding the Process of Economic Change*, Princeton, Princeton University Press, (trad. française: *Le Processus du Développement Économique*), Paris, Editions d'organization.

Padilla, M., Rastoin J.L. and Oberti, B. (2005), 'De Platon à Amartya Sen: le désordre alimentaire vu par les grands penseurs', in: G. Ghersi (dir.), *Nourrir 9 milliards d'hommes*, Paris, ADPF, pp. 52–55.

Pérez, R. (2004), *La gouvernance de l'entreprise*, Paris, Repères, La Découverte.

Raoult-Wack, A.L. (2001), *Dis-moi ce que tu manges*, Paris, Gallimard, coll. Découvertes.

Rastoin, J.L. (2005a), 'Un système alimentaire socialement responsable est-il un oxymore?', in: F. Le Roy and M. Marchesnay (dirs), *La responsabilité sociale de l'entreprise*, Editions EMS, Management et Société, chapter 12, pp. 157–168.

Rastoin, J.L. (2005b), 'Agriculture, alimentation, développement rural: quelle politique publique?', in: *Economies et Sociétés, Cahiers de l'ISMEA,* Tome XXXIX, n°5, mai 2005, série « Systèmes agroalimentaires », Paris, AG, n° 27, pp. 827–834.

Rémésy, C. (2005), *Que mangerons-nous demain*?, Paris, Odile Jacob.

Schumacher, E.F. (1973), *Small is Beautiful: A Study of Economics as if People Mattered*, London, Blond and Briggs.

Sen, A.K. (1981), *Poverty and Famines*, Oxford, Clarendon Press.

Smith, A., Watkiss, P., Tweddle, G., McKinnon, A., Browne, M., Hunt, A., Trevelen, C., Nash, C. and Cross, S. (2005), *The Validity of Food Miles as an Indicator of Sustainable Development*, Oxon, UK, AEA Technology Environment/DEPRA.

Watts, D., Ilbery, B. and Maye, D. (2005), 'Making Reconnections in Agro-Food Geography: Alternative Systems of Food Provision', *Progess in Human Geography*, vol. 29, 1 pp. 22–40.

Winter, M. (2003), 'Embeddedness, the New Food Economy and Defensive Localism', *Journal of Rural Studies*, vol. 19, pp. 23–32.

Chapter 3
Sustainability and Agri-Environmental Policy in the European Union: A Meta-Analytic Investigation

Katrin Oltmer, Peter Nijkamp, Raymond Florax and Floor Brouwer

Introduction

Land use and land cover have in recent years become major policy and research issues. In particular, as a result of deregulation and decentralization trends, the number of stakeholders involved with land use planning has increased, while the economic interests in land have also risen. Consequently, land plays a critical role not only in urban rehabilitation projects, real estate development and industrial site planning, but also in environmental management and agricultural land use policy, which is the focus of this chapter.

In the history of economic thought, varying attention has been given to land as an economic production factor. A dominant role, for instance, was assigned to land as a basic input to the creation of economic welfare in the period of the physiocrats. In the neoclassical world, land mainly assumed a functional economic position, as productivity and welfare differences between regions could be explained, inter alia, by different soil conditions (see also Giaoutzi and Nijkamp, 1994). More recently – partly as a result of the emergence of ecological economics – it has been recognized that land has not only a productive but also a consumptive meaning (e.g. as a recreational resource) in a sustainable development perspective (for a review, see van den Bergh, 1996). Furthermore, it is increasingly recognized that the condition of the soil has a variety of direct and indirect impacts on the quality and resilience of ecosystems with serious consequences for biodiversity, not only locally but also globally.

In the spirit of the debate on sustainable development, land use change has recently become a new focal point of interest for both scientists and policy makers, e.g. in relation to deforestation (Chomitz and Gray, 1996); soil rehabilitation (Beinat and Nijkamp, 1999; Nijkamp, 2000); or urban renewal (Finco and Nijkamp, 2000). Clearly, land use is a multifaceted phenomenon, driven by several economic, demographic, technological and physical factors (such as crop prices, population growth, harvest techniques, climatological factors, etc.). In the present chapter, we will mainly focus on land use in rural and agricultural areas. It is conceivable that a great variety of modelling approaches have been developed to investigate the dynamics in agricultural land use. In a recent study (Groeneveld and van Ierland, 2000), the authors distinguish and review the following types of modelling approaches: analytical models, optimization models, general equilibrium models, spatial equilibrium models, econometric models, heuristic decision models, empirical models and discrete choice models.

It goes without saying that the sustainability debate has prompted new challenges and research directions in agricultural land use research. For economists, the notion of sustainable development has meant a major new challenge, as they were forced to broaden conventional land use frameworks towards the domain of ecological systems or even international negotiation tables (for a review, see again van den Bergh, 1996). At present, several economic studies are still rather abstract and theoretical in nature (e.g. by seeking optimal trajectories or game-theoretic equilibria), but an increasing number of studies can be found which offer interesting applied work in the area of agricultural land use (see, e.g., Miller and Plantinga, 1999; Parks and Schorr, 1997).

The focus on local land use and sustainability conditions has also led to a rising interest in research which moves away from global sustainability analyses towards empirical policy-relevant research at the regional land use level (see, e.g., Giaoutzi and Nijkamp, 1994). This new interest in regional sustainability analysis is caused by several factors: a region is a properly demarcated area with some degree of homogeneity which allows for a more operational empirical investigation; a region is usually also subject to a properly regulated administrative competence and institutional control, so that there is more scope for a relevant policy analysis of sustainability issues; and finally, the statistical database at a regional level is often more appropriate for monitoring, analysing and modelling the economy and ecology of an area. Indeed, in recent international agreements and sustainability studies, the region has assumed a prominent position.

In Agenda 21, agreed upon at the Rio Summit, it was stated that land use planning should strive for '*promoting sustainable human settlement development*'. The fulfilment of such a task requires a clear analytical framework. In Figure 3.1 an illustrative presentation of the scope of sustainable land use planning is given. This figure clarifies that it is no surprise that in recent years land use has increasingly become a battlefield of conflicting interests (see also Frederick and Rosenberg, 1994).

Over the last few centuries, a significant and progressive transformation of natural areas into areas that support agricultural, urban or industrial functions has been observed. Apart from Europe, where both forests and grasslands show a slight expansion, the overall trend is towards a substantial loss of natural land in favour of cropland. The combined pressure of key factors such as population growth, food production, wood production and land tenure arrangements (Pearce, 1991) has affected as much as 40 per cent of the forests and grasslands of some areas. This trend will continue in the future, as the demand for space and natural resources will probably continue to rise. Irrigated land, cropland, rangeland and pasture will increase in absolute terms, but their availability per capita will also

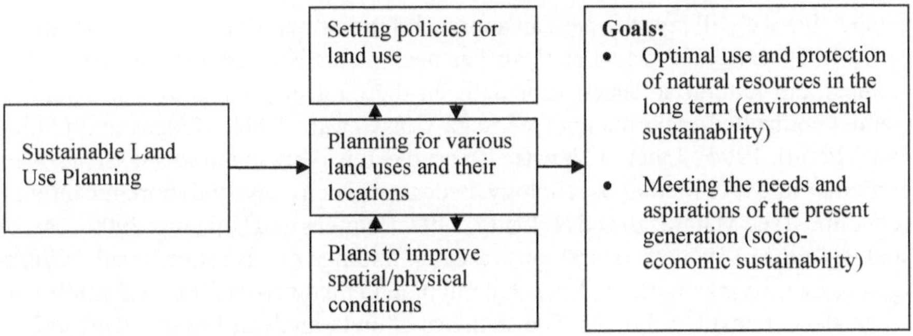

Figure 3.1 Sustainable planning of land use

Source: Van Lier et al. (1994).

decrease with rising population. Without countermeasures, this will necessarily lead to further pressure on land, to an increasing load on environmental quality, and to an impoverishment of natural resource capital. The negative effects of land-use exploitation are manifested in soil erosion, loss of habitats, increased vulnerability of the soil, a decrease in the carrying capacity of land, landscape modification and loss of natural amenities (see Beinat and Nijkamp, 1998). Therefore, it is no surprise that, over the years, a variety of agricultural land use policies have been developed with the aim to find a balance between economic efficiency and ecological quality.

Against the above background observations, the present chapter aims to offer a framework for the comparative analysis of agricultural land use practices in various European countries. In this chapter, agricultural land use practices are represented by what are called 'environmental driving force indicators': namely, the use of nitrogen fertilizer, livestock density, and grassland area. The main emphasis will be on the identification of drivers in agricultural land use practices by means of meta-analytic methods. Some of these drivers may stem from specific policy measures and others from general market or external conditions. Here, drivers stemming from specific policy measures and from the structure of the agricultural sector will be investigated.

This chapter considers two major research questions on the environmental aspects of agricultural land use. One is concerned with the assessment of environmental effectiveness of agri-environmental policies in the European Union, as a consequence of recent changes in agricultural and environmental policy. From the perspective and the need to draw lessons from comparative case study research in this field, the second research task of this paper deals with a related methodological issue: namely, whether meta-analysis is a suitable tool for the policy assessment of agri-environmental initiatives in the EU.

The various experiences of agricultural policy in various European countries call for a systematic research synthesis and comparison. From a methodological perspective we will employ here an approach for comparative case study research called 'meta-analysis'. Meta-analysis has become an established technique in the medical and natural sciences, especially in the case of comparative analysis of (semi-) controlled experiments (see, e.g., Glass et al., 1984; Hedges and Olkin, 1985; Petitti, 1994). Later on, it was also used extensively in the social sciences, in particular in experimental psychology, pedagogy, sociology, and more recently in economics (see Matarazzo and Nijkamp, 1997; Baaijens and Nijkamp, 2000). Meta-analysis aims to synthesize previous research findings or case study results with a view to identifying commonalities that might lend themselves for transferability to other, as yet unexplored cases. The statistics of meta-analysis has in the meantime become rather well developed. Especially where quantitative case study results are concerned, significant progress has been made. In this chapter, we will address in particular an ANOVA-type of meta-analysis adapted to effect size estimations in order to identify common drivers of agricultural dynamics in Europe.

The chapter is organized as follows. Section 3.1 describes agri-environmental policies in the EU, in particular the structure of the agri-environmental Regulation 2078/92 of the Common Agricultural Policy. Section 3.2 gives a short introduction to the use of environmental indicators in policy analysis and explains the environmental indicators used in our analysis. Section 3.3 presents the input data for the analysis that originate from the case studies of an EU project. The methodology of meta-analysis and the statistical procedures applied in our analysis are demonstrated in Section 3.4. Section 3.5 then reports the results, and, finally, Section 3.6 gives conclusions and recommendations.

3.1 Agri-environmental policies in the European Union

Along with the MacSharry reform of the European Union's Common Agricultural Policy (CAP) in 1992, three accompanying measures for stimulating the restructuring of the agricultural sector were introduced. One of them is called the 'agri-environmental' measure or, in formal terms, EC-Council Regulation 2078/92. This regulation is concerned with the implementation of special programmes that are intended to support and encourage farmers to introduce or continue with agricultural production methods consistent with the requirements of environmental protection and the maintenance of characteristic landscapes and the countryside. This implies that it is not only a framework for the stimulation of sustainable agriculture, but also for the multifunctionality and originality of rural space in Europe (Buller, 2000). The other two accompanying measures are the early retirement scheme for farmers (Regulation 2079/92) and the programme for the afforestation of agricultural land (Regulation 2080/92) (see Soufi and Tuddenham, 1995; Brouwer and van Berkum, 1996; Buller, 2000).

The agri-environmental Regulation 2078/92 is a co-financed instrument. This means that the Member States can apply for co-funding of up to 50 per cent and even up to 75 per cent for Objective 1 regions.[1] The financial source for the accompanying measures is the EAGGF (European Agricultural Guidance and Guarantee Fund), and the amount of money spent on them is rapidly increasing compared with traditional EAGGF expenditures such as those on the crop or the animal sector. However, the amount of money the EU spends on agri-environmental measures is still not more than 4 per cent of total CAP expenditures (Buller, 2000). The total amount of money spent on agri-environmental measures is obviously higher because, as mentioned above, the national governments also contribute to the financing of these measures. A special feature of Regulation 2078/92 is that, although the participation in agri-environmental programmes is voluntary for farmers, it is obligatory for the Member States to implement such programmes. It is thus the first common European framework for national policies in the agri-environmental field (Brouwer and Lowe, 1998).

Regulation 2078/92 is a very diverse and broad instrument that should be sufficiently flexible to consider the differences in geographical conditions, agricultural production systems and rural traditions within the territory of the European Union. Because of these diverging regional circumstances, it is obvious that the elaboration and implementation of Regulation 2078/92 takes place on a national, regional or even local level. As a result, Regulation 2078/92 includes about 2200 distinct measures incorporated in 127 programmes. 'Programmes' can be described as the way national or regional governments implement Regulation 2078/92, whereas 'measures' are the specific agri-environmental actions introduced at a local level as components of national or regional programmes (see Biehl, 1999). The European Commission has established a number of aid schemes that should be regarded by the Member States when applying for financial aid for these programmes. The aid schemes are described in Articles 2.1 and 2.2 of the Regulation, and they are shown in Table 3.1.

The table makes clear that the aid schemes do indeed comprise a wide range of agricultural practices promoting environmentally-friendlier ways of farming. Apart from this variety of agri-environmental measures, there are also different strategies concerning how to implement them. Buller (2000) distinguishes between four broad models of implementation. First, there are the targeted or zonal measures that aim at specific landscape types, natural regions or farming systems and at farmers located in a particular zone. Examples of this type of measure are the Environmentally Sensitive Area (ESA) schemes in Denmark and the United Kingdom.

Targeted or zonal measures are applied in most of the Member States. Secondly, there are wide horizontal schemes that cover whole nations or regions and which address certain eligibility criteria such as grassland in the 'Prime à l'herbe' in France. A third type of implementation strategy is a broad regulatory framework

1 Objective 1 regions are those whose development is lagging behind, in the sense that their per capita GDP is less than 75 per cent of the Community average over the past three years (EC, 1995–2000).

Table 3.1 Scheme objectives eligible for aid under Regulation 2078/92

Article 2.1 of Regulation 2078/92: Scheme objectives eligible for aid	
A	To reduce substantially the use of fertilizers and/or plant protection products, or to maintain the reductions already made; or to introduce or to continue with organic farming.
B	To change, by means other than those referred to in (a), to more extensive forms of crop, including forage production; or to maintain extensive production methods introduced in the past; or to convert arable land into extensive grassland.
C	To reduce the number of sheep and cattle per forage area.
D	To use other farming practices compatible with the requirements necessary to protect the environment and natural resources, as well as the maintenance of the countryside and landscape; or to rear animals or local breeds in danger of extinction.
E	To ensure the upkeep of abandoned farmland or woodlands.
F	To set aside farmland for at least 20 years with a view to using it for the purpose of enhancing the environment, in particular for the establishment of biotope reserves of natural parks or for the protection of hydrological systems.
G	To manage land for public access and leisure activities.
Article 2.2 of Regulation 2078/92	
Training and demonstration projects for farmers.	

Sources: CEC (1992); Buller (2000); Deblitz (1999).

that generally consists of a basic initial payment to participating farmers and a number of additional aid-schemes requiring further restrictions and accordingly higher payments. The Irish Rural Environmental Protection Scheme (REPS) is an example of this type of implementation strategy. Fourthly, there are measures that focus on specific actions, such as the conversion and maintenance of organic farming, the protection of local breeds in danger of extinction, or training and demonstrations projects for farmers. In general, it can be observed that schemes that demand changes in agricultural techniques involve higher payments than those focusing on the maintenance of existing extensive practices.

The EC has also proposed a categorization of the aid schemes into five groups: i) organic farming; ii) farming with environmental improvements; iii) maintenance of low intensity systems; iv) non-productive land management; v) training and demonstration projects. The preferences for these schemes amongst the Member States appear to vary significantly. For example, Mediterranean countries tend to use Regulation 2078/92 mainly for non-productive land management, which can be seen as a complementary source of income for farmers. Belgium, Denmark and Italy distinguish themselves from other Member States by allocating large proportions of their 2078/92 budget to organic farming, whereas Sweden and the Netherlands are in favour of training and demonstration projects (Buller, 2000).

By April 1997, 1.3 million contracts had been signed. This reflects around 18 per cent of farms and 17 per cent of the total Utilisable Agricultural Area (UAA) in the EU (Buller, 2000). The fact that the number of farms is slightly higher than the area under contract shows that there is a tendency of small-scale farmers to participate in the programmes.

It has to be mentioned that environmental concerns are not the only objective of Regulation 2078/92. Article 1 of the Regulation establishes three major goals: first, to accompany the changes to be introduced under the CAP reform in 1992; secondly, to contribute to the Community's policy objectives regarding agriculture and the environment; and thirdly, to contribute to providing an appropriate income for farmers (CEC, 1992). The *first goal* refers to the basic purposes of the 1992 CAP reform: namely, the reduction of overproduction and its increasing costs; the reduction of market support measures and the introduction of a system of direct payments. The *second goal* addresses the growing concern about the negative effects of agriculture on the environment, such as water pollution, biodiversity loss and landscape change. It is furthermore the first effort to comply with the Maastricht Treaty that requires EU environmental policy to be integrated into all other EU policies. The *third goal* is concerned with the maintenance and protection of extensive farming practices, not only against intensification but also against agricultural decline and withdrawal (Buller, 2000).

In the light of the global liberalization of agricultural trade, the last goal stated in Article 1 of Regulation 2078/92 is a particularly critical factor, since it can be interpreted as justifying the continuation of funding and subsidization of European agriculture disguised as 'green' CAP (Buller, 2000). Nevertheless, agri-environmental support payments to farmers are accepted according to the GATT agreement on agriculture. The WTO Member States agreed on a reduction of domestic support measures to agriculture by 20 per cent between 1995 and 2000 with respect to the support level in 1986–88. This reduction only refers to what are called 'Amber Box' support measures, which are those that have the most marked effect on agricultural production and hence also on trade. A typical example of an Amber Box support measure is price support, which gives farmers direct economic incentives to expand or reduce their production. Agri-environmental policies belong to 'Green Box' support measures. These measures are meant to have only a very small effect on production and trade, since the payments are supposed to be totally decoupled from production. Other kinds of Green Box measures are, for example, general services, such as research or pest and disease control, domestic food aid, and compensation payments for natural disasters. There are also 'Blue Box' support measures, which provide payments on the basis of a fixed amount of hectares or livestock in the frame of production limiting-programmes (Silvis and Rijswick, 1999).

There are still more criticisms about the current structure of agri-environmental policy in the EU. The most important one for policy makers is that, because of the wide variety of implementation strategies, it is rather difficult to carry out cross-national comparisons of scheme effectiveness and to evaluate the economic efficiency of the schemes in general. Another criticism is that the environmental policy target is in many cases far too broad and not adequately identified, so that potential positive effects on the environment cannot be evaluated. Furthermore, it is argued that Regulation 2078/92 is poorly integrated with other CAP policies. For instance, the maize premium of the regular CAP programme is in many cases higher than the grassland premium under the agri-environmental policy (Buller, 2000).

Being aware of the critical points of agri-environmental policy making, the EC has shifted the importance more and more towards the environmental objective, the second goal of Regulation 2078/92. In order to achieve these changes legally, in 1999 the EC introduced a new tool, viz. the *Integrated Rural Development Regulation* (Regulation 1257/99). This new regulation integrates not only Regulation 2078/92 but also other rural measures such as the Less Favoured Area scheme. In this new regulation, income support to farmers is no longer mentioned, and environmental goals are clearly specified for farmers who want to participate in agri-environmental policy programmes (Lowe and Baldock, 2000).

3.2 Environmental indicators

For a proper quantitative policy assessment, we have to resort to reliable indicators. The OECD (1997) defined three major functions of environmental indicators in agriculture: 1) they should provide information to policy makers and the general public about the state of the environment influenced by agriculture; 2) they have to help policy makers to better understand the cause-effect loops between agricultural activity and the environment; and 3) they have to assist in the evaluation of the effectiveness of agri-environmental policy instruments. In order to comply with these three demands, the OECD has proposed to apply what is called a Driving force-State-Response (DSR) framework. 'Driving forces' are the factors that cause environmental conditions to change. These factors include input and output levels of farm production, agricultural land use, and also natural processes and meteorological conditions. 'State' describes the actual condition of the environment, for instance, the nutrient level in groundwater and surface water or the number of protected species in a certain area. 'Response' refers to the reactions of policy makers and groups in society to the state of the environment.

Although the actual state of the environment would be the most appropriate indicator for policy evaluation, it is, especially in agriculture, also the most difficult one to assess. There are has several reasons for this. One of the most important is the time and space dimension inherent to the cause-effect loop between agricultural production and the state of the environment. This means that the effects of agricultural pollution might become visible only after a number of years, or that they spread out over long distances through, for example, water or air (Deblitz, 1999). Another important reason is that the assessment of state indicators is in most cases rather costly.

The most appropriate alternative is to take the driving force indicator as a measure for the effectiveness of agri-environmental policy. In this case, the driving force indicators are agricultural practices that have a certain effect on environmental quality. The indicators used in this study are the same as those used in the FAIR research project (see Section 3.4), since that project provides the data as input for the meta-analysis carried out in our study. The FAIR research project developed 12 different indicators based on nine particular agricultural practices. The agricultural practices were selected according to three criteria, viz. relevance, reliability, and realizability. Relevance implies the correspondence of the agricultural practices to the specific goal and actions of Regulation 2078/92. Reliability requires that the impact of the agricultural practices on the environment must be well known and scientifically proved. Finally, realizability refers to the availability of the appropriate data (Andersen et al., 1999).

For the purpose of meta-analysis a significant minimum amount of systematic and common data is needed. Since not all of the 12 indicators comply with this requirement, we were forced to employ only three of them: namely, mineral nitrogen fertilizer, livestock density, and grassland area per utilizable agricultural area. Our choice of indicators is hence solely based on data availability.

The actual relationship between the agricultural practices serving as our indicators and environmental quality is described in several scientific studies. Andersen et al. (1999) give a concise overview of the literature on these relationships. A short summary of this literature overview is given below.

- *Mineral nitrogen fertilizer* The excessive use of N-fertilizer can change the botanical composition of grassland by favouring particular species against others. This in turn harmfully influences specific bird populations that use grassland as their breeding and feeding habitat. Furthermore, intensive mineral N-fertilization increases the nitrogen stock in the soil, which results in a rate of nitrification that is higher than the nitrogen demand of the current crop. As a consequence, the surplus of nitrogen will leach into groundwater. In order to comply with European standards for drinking water, the level of nitrate in groundwater must not exceed 50 mg/litre. The mineral N-fertilizer indicator is measured in kg N-fertilizer per hectare, and it has a negative relationship with the state of the environment. A decreasing value of the indicator is therefore preferable.
- *Livestock density* A large number of livestock per agricultural area is equivalent to high levels of the manure and slurry on this area. This, in turn, is directly related to leaching of nitrate into groundwater resources. However, the actual relationship between livestock density is found out to be bell-shaped. This means that livestock densities that are either too high or too low result in a degradation of the traditional ecological system. In our case, the second half of the bell-shaped curve is of importance, as it implies that livestock densities have to be reduced in order to improve environmental quality. The livestock density indicator is measured in total livestock units (LU) per hectare of utilizable agricultural area. It has a negative relationship with the state of the environment, which means that decreasing livestock density is (in general) favourable for the environment.

- *Grassland area per utilizable agricultural area (UAA)* In comparison with arable land, grassland has many environmental advantages. First of all, the loss of nitrogen under grassland is significantly smaller than under arable land. Since ploughing accelerates the mobilization of nitrate, it is advantageous to prevent the conversion of grassland to arable land. Furthermore, the maintenance of extensive grassland is desirable because not only intensification but also abandonment negatively affects the variety of faunal and floristic species of grassland, which again has an unfavourable impact on grassland birds. Finally, conserving grassland is an ideal measure for the prevention of soil erosion through wind and water. From all these facts it can be concluded that there is a direct positive relationship between the share of grassland and the state of the environment, which means that an increase in the indicator is beneficial to the environment. The grassland indicator is measured as the percentage of grassland per UAA.

3.3 Input data: Case studies of an EU project

The case studies used in our empirical meta-analysis are the results of a 3-year project[2] funded by the European Union about the implementation and effectiveness of agri-environmental schemes established under Regulation 2078/92 (for the full project report see Schramek et al., 1999). The project includes nine EU countries: namely, Sweden, Denmark, Germany, Great Britain, France, Austria, Spain, Portugal and Greece. Additionally, it considers Switzerland in order to compare the experiences of non-EU-members that apply agri-environmental policies similarly to those of the EU. The research group was characterized by multidisciplinarity and consisted of agricultural economists, general economists, ecologists, geographers, landscape planners and sociologists.

The data collection took place through farm surveys based on a uniform questionnaire, 22 case-study areas were selected, two in each country, except for Sweden where four case-study areas were selected. In total, 1000 farmers were interviewed, 50 in each case study area (and 25 in the Swedish case-study areas). The study areas cover a wide range of European landscape types and different agri-environmental programmes, and are selected according to a limited number of agri-environmental issues, such as contamination of groundwater and soil, or biodiversity.

The objective of this research project was '...*to develop common and appropriately regionalized operational methodologies, and to apply these methodologies in order to analyse the implementation and effectiveness of EU-agri-environmental schemes established under Regulation 2078/92*' (Schramek et al., 1999, p.1). With the help of the questionnaire, the research group was not only able to identify and analyse farmers' participation in, and attitudes towards, agri-

2 Project FAIR 1 CT95-274 (Biehl, 1999).

environmental policies, but they were also able to trace the environmental and socio-economic impacts of EU policies. For the purposes of this chapter, we will mainly focus on the results of the environmental impact analysis.

The input data for the meta-analysis stem from the research group's analysis of the environmental effects of Regulation 2078/92 that was carried out on a case-study level (for a detailed description of this analysis, we refer to Andersen et al., 1999). As mentioned in the previous section, in the project 12 environmental indicators were proposed and developed on the basis of certain agricultural practices, of which three are selected for the analysis in the present chapter. The reasons for the limited availability of data from the other 9 indicators are that not all indicators are applicable to all case-study areas, and that response rates were too low in some areas to ensure reliable inferences. The indicators for the agricultural practices '*reduction in the use of mineral N-fertilizer (kg/ha)*', '*reduction of livestock density (LU/ha*' and '*increase of grassland area with respect to total agricultural area (% grassland/UAA*' are reflected by the average change rates per case study area of these practices over a 5-year period (1993–1997). The data for the calculation of these average change rates are taken from the individual farm questionnaires. The farmers interviewed are classified into two groups. On the one hand, there are farmers who are eligible for, and participating in, agri-environmental programmes, and, on the other hand, there are farmers who are also eligible but not participating. The approach of comparing the behaviour of participating farmers with that of non-participating farmers makes it possible to directly identify the environmental impact of the programmes concerned. In the research process of the FAIR project, the average change rates of the two groups of farmers were compared statistically on a case-study area level. The statistical test methods used for the comparison of the two groups are the t-Student test and the U-Mann Witney test.

The subdivision of the interviewees into participants and non-participants can be interpreted as a quasi-experimental research design. In this case, participating farmers act as the experimental group and non-participating farmers as the control group, albeit that both are eligible. The non-random assignment of subjects, on the basis of self-selection, may cause some bias in the analysis, but since there was no other way of creating a database, this shortcoming has to be accepted. It means that our results have to be interpreted with caution. The structure of the experimental and the control group is a proper base for conducting a meta-analysis, where what are referred to as 'effect sizes' are calculated. These reflect the relative difference between these two groups. Further explanations about meta-analysis as it is carried out in the present paper are offered in Section 3.5. The comparisons between the selected indicators' average change rates of participants and non-participants on a case-study level carried out by the FAIR research team gave the following results.

- *Kg N-fertilizer per hectare* This indicator appears to be relevant for most of the 22 case-study areas. However, lack of data meant that only nine case study areas could be analysed: i) Great Britain-Cambrian Mountains; ii) Great Britain-Devon Countryside; iii) Germany-Rhoen; iv) Germany-Wetterau; v) Denmark-Viborg County; vi) Denmark-Vestsjaelland; vii) Spain-Sahagún; viii) Austria-Nordburgenland; ix) Greece-Larisa. Significant and expected results – meaning that the average change rates in using N-fertilizer of participants are negative and significantly higher than those of non-participants – are found in Spain-Sahagún and Greece-Larisa at the 5 per cent level, and in Germany-Wetterau and Great Britain-Cambrian Mountains at the 10 per cent level. In all the other case-study areas the results are insignificant. An unexpected positive (but insignificant) average change rate of participating farmers is found in Denmark-Vestsjaelland.

- *Livestock units per hectare* This indicator is relevant for 14 case-study areas, of which two had to be dropped because of limited data availability, and because of a significant change in the reference area. The following 12 case-study areas were ultimately included in the analysis: i) Great Britain-Cambrian Mountains; ii) Great Britain-Devon Countryside; iii) Denmark-Viborg County; iv) Denmark-Vestsjaelland; v) Portugal-Moura; vi) Portugal-Castro Verde; vii) Austria-Nordburgenland; viii) Austria-Osttirol; ix) Switzerland-Schwarzwasser; x) Switzerland-Erlach/Seeland; xi) Germany-Rhoen; xii) France-Bocage-Avesnois. The only significant result – meaning that the average change rate of participants is negative and significantly higher than that of non-participants – is found in Germany-Rhoen at the 5 per cent level. In all other case-study areas, the results are insignificant. Unexpectedly positive (but insignificant) average change rates of participating farmers are found in Switzerland-Schwarzwasser, Denmark-Vestsjaelland, Austria-Nordburgenland, and Austria-Osttirol.
- *Grassland area (% per UAA)* This indicator concerns 18 case-study areas. However, only 13 case-study areas could be analysed, again because of limited data availability and significant changes in the reference area. The analysed case-study areas are: i) Great Britain-Cambrian Mountains; ii) Great Britain-Devon Countryside; iii) Denmark-Viborg County; iv) Denmark-Vestsjaelland; v) Sweden-Enkoping; vi) Sweden-Offerdal; vii) Sweden-Vallakra; viii) Austria-Nordburgenland; ix) Austria-Osttirol; x) Switzerland-Schwarzwasser; xi) Switzerland-Erlach/Seeland; xii) Germany-Rhoen; xiii) France-Bocage-Avesnois. The only significant result with an expected positive average change rate that is higher for participants than for non-participants is found in Sweden-Enkoping at the 10 per cent level. Another significant but paradoxical result: namely, that the average change rate of participants is negative and higher than that of non-participants, is found in Great Britain-Devon Countryside. In all other case-study areas the results are insignificant. In Switzerland-Schwarzwasser, France-Bocage-Avesnois, Great Britain-Cambrian Mountains, Sweden-Offerdal and Sweden-Vallakra the average change rates are unexpectedly negative (but insignificant).

To summarize, the number of expected and significant results of the change rates for the three indicators is rather limited. At the 5 per cent level, 2 out of 9 results are significant for N-fertilizer, 1 out of 12 results is significant for livestock density, and for grassland no significant results were found at all. However, it has to be kept in mind that the sample sizes in the individual case study areas tend to be rather small, which increases the probability of accepting the null-hypothesis although it may be false. This problem will be further elaborated in the following section. 9, 12 and 13 observations, respectively, for the three indicators N-fertilizer, livestock density, and grassland area are available for the analysis in this chapter. More observations per indicator would certainly improve and strengthen our analysis because they would allow a more varied and differentiated investigation. However, since we are confined to the limited observations available, the analysis may be seen as a first exploration to apply the techniques of meta-analysis to agri-environmental policy evaluation. The statistical procedure of the meta-analysis employed in this paper is described in the following section.

3.4 Methodology of research synthesis

3.4.1 Introduction

The methodology adopted for our empirical case study is based on meta-analysis. Meta-analysis has already a quite remarkable history in psychology and medical science and found only recently found its way to regional and environmental economics. The development of meta-analysis in psychology and medical science for the main part is related to large numbers of case studies on the same scientific issue performed in an experimental and largely standardized context, which forms a perfect base for statistically-based meta-analysis. The lack of experimental and standardized conditions in many fields of the social sciences (including economics) is in fact the major criticism of applying meta-analysis to social science issues. In order to be able to compare existing research results in a strict statistical way, studies should involve quantitative factors and identical units, or at least results that can be transformed into some common unit or index (van den Bergh and Button, 1997).

Because of their quasi-experimental approach, the results from the case studies carried out in the FAIR project form, for environmental economic purposes, suitable inputs for meta-analytic research. At this point, the potential additional value of applying meta-analysis to these case studies has to be identified.

The previous section presented the results of the analysis of the average change rates in the individual case-study areas of the FAIR project, which in many cases show insignificant results. It was mentioned that this could be caused by the relatively small sample sizes. Standard statistical theory would tell us that parameter estimates from large sample sizes are more robust than those from small sample sizes, because the variance around parameter estimates from large samples is smaller (Shadish and Haddock, 1994).

Consequently, estimates from large samples tend to have more significance. On the other hand, estimates obtained from rather small samples are, because of their larger variances, subject to the risk of Type II errors, which means accepting the null hypothesis, even though it may be false (Hunter and Schmidt, 1990). This problem is aggravated if the estimated population effect is small. Summarizing case-study results from small samples by simple vote-counting procedures, which mean counting significant results only, might lead to the conclusion that the average effect of the intervention is not significantly different from zero (Hedges and Olkin, 1985). Meta-analysis artificially increases the sample size by pooling, so to speak, all sample sizes from the individual studies for the calculation of the average effect size. An 'effect size' is a generic term that refers to the magnitude of an effect or, more generally, the size of the relation between two variables (Cooper and Hedges, 1994). (A detailed description of effect sizes is given in Section 3.4.2). Hence, meta-analysis does not take into account the significance level of the individual studies, but only their effect sizes. Because of the increased sample size, the calculated average effect size is generally significantly different from zero (Ijskes, 1999). In our case, we will test if the average change rates of participating and non-participating farmers with respect to the three indicators are indeed significantly different from each other, even though most of the original results are insignificant.

Another question that meta-analytic techniques are able to answer is whether individual studies share a common effect size, or, in other words, whether there is a single overall effect size that describes the general magnitude of the intervention. If this is not the case, meaning that the individual effect sizes are too heterogeneous to support the hypothesis of a common effect size, there must be factors at work that are responsible for the variations among the individual effect sizes. The identification of these factors is another task of the meta-analysis carried out in this paper.

3.4.2 The effect size

Several different forms of effect sizes can be found in the current meta-analytical literature. In the medical, social and psychological sciences, the areas that are considered to be the traditional disciplines in the application of meta-analysis, two types of effect sizes are most commonly used: the d-type, and the r-type. The most famous effect sizes of the d-type are Hedges' g, Cohen's d and Glass's Δ. An example of an r-type effect size is the correlation coefficient r. Effect sizes of the d-type are all standardized mean differences of control and experimental groups, which differ from each other with respect to the way of standardization. Hedges' g uses the pooled (experimental *plus* control group) standard deviation that is calculated with degrees of freedom, which are the total number of observations minus the number of groups, for standardization. Cohen's d also standardizes by means of the pooled standard deviation but uses the total number of observations instead of degrees of freedom for the computation. Finally, Glass's Δ uses only the standard deviation of the control group for standardization (Rosenthal, 1991, 1994).

The choice of which effect size to apply depends partly on the type of data available, but also on personal preferences. In our case, the original studies compare two groups. They also report the means, standard errors and sample sizes of these groups, so that it is most appropriate to calculate an effect size of the d-type. This analysis employs Hedges' g as its effect size. Effect sizes of the d-type and of the r-type are actually convertible to each other. The correlation coefficient, r, is hence just another way of interpreting the effect sizes of the d-type (Hedges and Olkin, 1985). An interpretation of effect sizes of the d-type is the following.

An effect size of the d-type reflects the difference between an experimental and a control group in such a way that it is independent of sample size and unit of measurement. In fact, the effect size gives the difference between an experimental and control group in standard deviation units (Rosenberg et al., 1997). Hedges and Olkin (1985) interpret the effect size as the z-score of the normal cumulative distribution function, where its respective $\Phi(z)$-value is the proportion of control group scores that is less than the average score of the experimental group. For example, an effect size of 0.3 signifies that the score of an average individual of the experimental group exceeds the score of 62 per cent ($\Phi(0.3) = 0.62$) of the individuals of the control group. Rosenberg et al. (1997) give Cohen's convenient rule of thumb about the interpretation of effect sizes. This rule says that 0.2 implies a small effect, 0.5 a medium effect, and 0.8 a large effect. Everything above 1.0 is considered to be a very large effect.

3.4.3 Meta-analysis in four steps

The meta-analysis performed in this paper is divided into four steps. Step 1 is the calculation of effect sizes for each case study area with respect to the selected environmental indicators. Step 2 is the combination of these effect sizes for each environmental indicator. In Step 3 it has to be investigated whether the estimated effect sizes are homogeneous, which means whether the effect sizes from the individual case studies share a common effect size. This is done by testing the null-hypothesis that there is no variation among the effect sizes. If this test is rejected, Step 4 has to be carried out, and that is the moderator analysis. The description of the statistical procedure is based on Hedges and Olkin (1985), Rosenthal (1991, 1994), and Shadish and Haddock (1994).

Step 1: Calculation of the effect size As mentioned above, this analysis employs Hedges' g as its effect size. Hedges' g is calculated according to the following formula:

$$g = \frac{M_E - M_C}{S_p}, \tag{3.1}$$

where M_E is the mean of experimental group, and M_C the mean of the control group. S_p is the pooled sample standard deviation computed as:

$$S_p = \sqrt{V_p} = \sqrt{\frac{(N_E - 1)V_E + (N_C - 1)V_C}{N_E + N_C - 2}}, \tag{3.2}$$

where V_E and V_C are the variances of the experimental and control group and N_E and N_C the experimental and control group sample sizes, respectively.

Rosenthal (1991, 1994) and Hedges and Olkin (1985) point out that g is negatively biased, especially when sample sizes are small and population effects are large. Because of the small sample argument, our analysis uses the adjusted, unbiased g, viz. g^u, that is obtained by applying:

$$g^u = g * c(m), \tag{3.3}$$

where

$$c(m) = 1 - \frac{3}{4(m) - 1}, \tag{3.4}$$

and m are the degrees of freedom computed from the experimental and control group (N_E+N_C–2). However, in our analysis the actual difference between g and g^u turns out to be rather small. In our calculations, c(m) lies around 0.98, which is close to 1 and hence almost negligible.

Step 2: Combining effect sizes It was already noted above that larger samples produce more significant and reliable estimates. It is hence suitable to weight the effect sizes of large sample studies more heavily before combining them. According to Shadish and Haddock (1994), the most appropriate weight is the inverse of the variance of the respective effect sizes, as shown in the following formula:

$$w_i = \frac{1}{v_i}, \tag{3.5}$$

where w_i is the weight, and v_i the variance of the i-th effect size calculated according to formula (3.2).

The combination of the different Hedges' g's obtained from *k* case studies gives the average effect size, $\overline{G_\bullet}$ that is calculated as:

$$\overline{G_\bullet} = \frac{\sum_{i=1}^{k} w_i g_i}{\sum_{i=1}^{k} w_i}. \tag{3.6}$$

In order to find out whether the average effect size is significantly different from zero, or, in other words, whether there is a significantly positive or negative average effect due to the intervention, the confidence interval of the average effect size has to be calculated. This requires the calculation of the standard error of the average effect size, $s_{\bullet}$, that is given by:

$$s_{\bullet} = \sqrt{v_{\bullet}} = \sqrt{\frac{1}{\sum_{i=1}^{k} w_i}}, \tag{3.7}$$

where $v_{\bullet}$ is the variance of the average effect size. Subsequently, the confidence interval can be computed according to:

$$\overline{G_{\bullet}} \pm C_{\alpha} * s_{\bullet}, \tag{3.8}$$

where C_{α} is the critical value of the standard normal distribution.

Alternatively, the null hypothesis that there is no average effect can be tested with the Z-statistic, and that is:

$$Z = \frac{\left|\overline{G_{\bullet}}\right|}{s_{\bullet}}. \tag{3.9}$$

If Z exceeds 1.96, the 95 per cent two-tailed critical value of the standard normal distribution, the null hypothesis can be rejected, and it can be concluded that the intervention has a significant average effect.

Step 3: Test on homogeneity of effect sizes Equation (3.6) assumes that all individual studies share a common effect size. This is certainly a very strong assumption and in most cases this is actually not the case. The test to ascertain which of the effect sizes from all individual studies are indeed not homogeneous is called the Q-test, and is represented by the following formula:

$$Q = \sum_{i=1}^{k} \frac{\left(g_i - \overline{G_{\bullet}}\right)^2}{v_i}. \tag{3.10}$$

If the value of Q exceeds the upper tail critical value of the χ^2-square distribution with k-1 degrees of freedom, it has to be assumed that the effect sizes of the individual studies are not homogeneous, and that the individual studies do not share a common effect size. $\overline{G_{\bullet}}$, as calculated in equation (3.6), has therefore to be interpreted as the mean of the observed effect sizes and not as a single effect parameter.

The heterogeneity of the effect sizes of the individual studies shows that there must be factors influencing the magnitude of the effect sizes. These factors are called 'moderator variables'. The analysis of moderator variables is described in the next step.

Step 4: Analysis of moderator variables Moderator variables are the factors that determine the variations in the effect sizes among the individual studies. Another interpretation of moderator variables is that they are able to identify important study characteristics. In our case, moderators should explain the variations of the policy effect in the different case-study areas. In other words, they should reflect the reasons why in some case-study areas there is a greater difference in behaviour between participating and non-participating farmers with regard to a particular indicator than in other case-study areas.

In general, moderator variables can be roughly categorized into three groups. First, there are moderators based on the underlying theoretical framework. In our case, an example of a moderator of the first type would be the premium level. Theoretically, it can be assumed that higher premium levels would induce larger changes in behaviour with respect to the particular agricultural practice indicators. Secondly, there is the group of moderators including variables that reflect the setting of the particular case study. Variables that reflect the setting of a case study describe country or time-specific characteristics. Thirdly, there is the group of moderators that refer to methodological issues of the primary case studies. These variables represent the way in which the analysis in the primary study is carried out. Examples are the statistical method used, the functional form chosen, or the type of data employed in the primary study. In the present chapter, the individual case studies will all apply the same statistical technique. This means that methodological moderators are assumed not to be very important in our case.

Certainly, the list of potential moderator variables is very long, and again the availability of information is the determining factor of which moderator variables to choose. The analysis in this current chapter tests the existence and importance of the following moderator variables.

1. *Average premium per hectare* Theoretically, higher premiums would imply that farmers would be more stimulated to change their behaviour with respect to the relevant agricultural practice indicators. Therefore, higher premiums would be related to larger effect sizes. The FAIR project reports average premiums per farm and average farm sizes of participating farmers for all case-study areas. The moderator variable *average premium per hectare* is calculated by dividing average premium per farm by average farm size of participating farmers for all relevant case-study areas. With this moderator it is tested whether higher premiums do indeed result in higher effect sizes.

2. *Average farm size* Larger farms are supposed to be more innovative and more creative with respect to attaining alternative sources of income. Large farms have more hectares at their disposal. This means that they have a higher probability that some of their land is located in an eligible area or that some of their land is marginal anyway. (Total agri-environmental payments per farm will be higher for large farms, meaning that it is more attractive for them to sign a contract). With this moderator we want to test if the variable farm size has an effect on the magnitude of the effect size.
3. *Absolute level of indicator in 1997* Case-study areas that have in general a relatively low (for N-fertilizer and livestock density) or high (for grassland) level of the indicator might have lower change rates of participating farmers and hence lower effect sizes. (The level of the indicator for the starting year (1993) would hence be more suitable. Unfortunately, this is not reported in the FAIR study.)

In the FAIR project, all case-study areas are categorized into four groups, each of them describing the characteristics of the agricultural production structure in that area. The four different categories are: intensive arable farming; extensive arable farming; intensive husbandry farming; and extensive husbandry farming. Unfortunately, the number of observations available to us was not large enough to use this differentiated categorization in one moderator analysis. Therefore, we had to simplify this categorization into the moderators 'intensive versus extensive farming', and 'arable versus husbandry farming' and perform two separate analyses on these two moderator variables. The moderators 'intensive versus extensive farming' and 'arable versus husbandry farming' are only tested for the indicator Nitrogen-fertilizer.

4. *Intensive versus extensive farming* With this moderator it is tested whether effect sizes in areas of intensive farming differ significantly from those in areas of extensive farming.
5. *Arable versus husbandry farming* With this moderator it is tested whether effect sizes in areas of arable farming differ significantly from those in areas of husbandry farming.

The most basic way to perform a moderator analysis is as follows. First of all, the sample of effect sizes has to be subdivided into two (or more, depending on the number of observations) groups that are associated with a particular characteristic reflected by a moderator variable. Subsequently, a meta-analysis, as described in Steps 1 to 3 has to be performed on the separate groups. Additionally, two more Q-tests have to be carried out. Firstly, there is the Q-test on heterogeneity between the groups: the Q-between test. Secondly, there is the Q-test on heterogeneity within the groups: the Q-within test. The Q-between statistic tests the null hypothesis that there is no variation across the group mean effect sizes. In other words, it tests whether a particular moderator variable does indeed have a significant influence on the effect size. The Q-between statistic is given by the following formula:

$$Q_{between} = \sum_{i=1}^{p} \frac{\left(\overline{g_{i\bullet}} - \overline{G_{\bullet}}\right)^2}{v_{i\bullet}}, \tag{3.11}$$

where p is the number of groups, $\overline{g_{i\bullet}}$ the average effect size of the *i*th group, $\overline{G_{\bullet}}$ the overall average effect size (formula (5.6), also called the grand weighted mean), and $v_{i\bullet}$ the variance of $\overline{g_{i\bullet}}$, calculated according to formula (3.7), taking into account only the observations in that particular group.

The Q-within statistic is presented by the following formula:

$$Q_{within} = \sum_{i=1}^{p} \sum_{i=1}^{m} \frac{\left(g_{ij} - \overline{g_{i\bullet}}\right)^2}{v_{ij}}, \tag{3.12}$$

where m is the number of observations in the *i*th group, g_{ij} the *j*th effect size in the *i*th group, and v_{ij} its variance, according to formula (3.7), taking into account only the observations in that particular group. In fact, the Q-within statistic is the sum of the Q-tests (formula (3.10)) applied to every single group:

$$Q_{within} = Q_{w1} + Q_{w2} + ... + Q_{wp}. \tag{3.13}$$

The sum of the Q-between and the Q-within statistic results in the overall Q-test applied to all observations (formula (3.10)):

$$Q = Q_{within} + Q_{between}. \tag{3.14}$$

In an ideal case, the selected moderator variable explains total heterogeneity such that most of the heterogeneity is between groups. If there is still heterogeneity within groups, the selected moderator variable is not able to explain all the variation between the effect sizes. If the number of observations within the groups is still large enough, a moderator analysis can be performed within the groups. This procedure could continue until there is no within-group heterogeneity anymore.

3.5 Results of the effect size analysis

This section presents the results of the meta-analysis applied to the evaluation of the three agri-environmental indicators: N-fertilizer; livestock density; and grassland area. Section 3.4.1 describes the outcomes of Steps 2 and 3. Section 3.5.2 gives the results of the moderator analyses.

3.5.1 Combined effect sizes and homogeneity test

The outcomes of Step 2 (combining effect sizes) and Step 3 (test on homogeneity) as described in the previous section are reported in Table 3.2.

The table shows that the combined effect sizes of all three indicators are significantly different from zero (the confidence intervals do not include zero). Here, the effect of the increased sample sizes becomes visible. Although most of the original case studies show insignificant results, the combined effect sizes demonstrate that there is an overall difference between the change rates of participating and non-participating farmers.

Table 3.2 Results of Step 2 and Step 3

	k	N (N_E, N_C)	Hedges' g	Var	SE	conf. interval Min	conf. interval Max	Q	$P_{(Q)}$
N-fertilizer	9	349 (242/107)	-1.573	0.236	0.154	-1.874	-1.272	52.24	0.000
Livestock	13	630 (445/185)	-0.816	0.012	0.111	-1.033	-0.598	161.81	0.000
Grassland	13	569	-0.831	0.015	0.122	-1.07	-0.591	169.84	0.000

Notes: k: number of case study areas; N: number of individual farmers; N_E: number of individual farmers in experimental group (participants); N_C: number of individual farmers in control group (non-participants); VAR: variance of Hedges' g; SE: standard error of Hedges' g.

The effect sizes of the indicators N-fertilizer and livestock density have the expected negative sign. However, the sign of the effect size of the indicator grassland is unexpectedly negative. This result is paradoxical because the policy is meant to increase the area of grassland. The fact that the confidence interval does not include zero makes this result even more contradictory.

The indicator N-fertilizer has the highest average effect size: namely, -1.57. According to the interpretation of effect sizes described in Section 3.4.2, 94 per cent ($\phi(1.57) = 0.94$) of the change rates of non-participating farmers are lower than the average change rate of participating farmers. Applying Cohen's rule of thumb (see Section 3.2), it can be stated that an effect size of -1.57 reflects a very large effect of the policy intervention regarding the use of fertilizer. It has to be noted that the effect sizes do not say anything about the difference in the actual size of the change rates of participating and non-participating farmers but only about the percentage value at which the change rates of non-participants lie under the average change rate of participants.

The effect size for the indicator livestock density is -0.82. This means that 79 per cent of the change rates of non-participating farmers are lower than the average change rate of participating farmers. According to Cohen's rule of thumb, this effect size exhibits a large effect of the policy intervention with regard to livestock density.

However, the Q-test on homogeneity signifies at a very high significance level for all three indictors that the effect sizes of the individual case-study areas are heterogeneous. This means that the case-study areas do not share a common effect size, and that the calculated effect size is only the mean of the effect sizes in the individual case-study areas.

3.5.2 Moderator analyses

Since the calculated effect sizes do not pass the Q-test on homogeneity, a moderator analysis as described in Step 4 has to be carried out. First, the moderator '*average premium per hectare*', secondly, the moderator '*average farm size of participating farmers*', and, thirdly, the moderator '*average absolute value in 1997*' (of the indicator) will be tested. Finally, the moderators '*intensive versus extensive farming*' and '*arable versus husbandry farming*' will be considered, but only for the indicator N-fertilizer.

I) Average premium per hectare The results of the moderator analysis '*average premium per hectare*' are shown in the following table.

Table 3.3 Results of moderator analysis 'average premium per hectare'

2-groups	**N-Fertilizer**			**Livestock Density**			**Grassland**		
Analysis	**Hedges' g**	**Q**	$P_{(Q)}$	**Hedges' g**	**Q**	$P_{(Q)}$	**Hedges' g**	**Q**	$P_{(Q)}$
< 40 ECU	-1.31	22.17	0.00	-0.51	106.17	0.00	-0.81	66.06	0.00
> 40 ECU	-1.83	27.19	0.00	-1.78	31.83	0.00	-0.85	103.76	0.00
Q between		2.88	**0.09**		23.81	0.00		0.02	**0.88**
Q within		49.36	0.00		138	0.00		169.82	0.00
3-groups Analysis									
< 30 ECU	-0.80	8.24	0.02	-0.41	38.47	0.00	-0.64	63.24	0.00
> 30 ECU	-2.54	0.45	**0.8**	-0.60	69.15	0.00	-1.46	9.229	0.03
> 100 ECU	-1.23	18.13	0.00	-1.92	24.35	0.00	-0.26	81.45	0.00
Q-between		25.42	0.00		29.84	0.00		15.93	0.00
Q-within		26.82	0.00		131.97	0.00		153.91	0.00

For the moderator '*average premium per hectare*', two kinds of analyses were carried out. In the first analysis, the effect sizes are divided into two groups. The first group comprises all case-study areas where the average premium is less than 40 ECU per hectare, and the second group includes all case-study areas where the average premium is greater than 40 ECU per hectare. For the indicators N-fertilizer and livestock density, the results are as expected: namely, that higher average premium per hectare result in higher effect sizes. (Recall that a higher effect size does not indicate higher actual change rates of the indicators for participating farmers, but only that a higher percentage of the change rates of non-participating farmers is lower than the average change rate of participating farmers.) The Q-between tests are highly significant for livestock density and significant at the 10 per cent level for N-fertilizer. This means that the effect sizes of the two groups are significantly different from each other. However, the Q-within statistics still indicate heterogeneity among the effect sizes in the two groups. For the indicator grassland, the effect sizes of the two groups are not significantly different from each other (the Q-between test cannot reject the null hypothesis of homogeneity).

Since the Q-within test in the 2-groups analysis still indicates heterogeneity among effect sizes; a second analysis was carried out. In this second analysis, we tried to find out whether a division into three groups might improve the Q-within tests. The first group includes all case-study areas where the average premium is less than 30 ECU per hectare; the second group contains all case-study areas with an average premium between 30 ECU and 100 ECU per hectare; and the third group comprises all case-study areas where the average premium per farm is above 100 ECU per hectare. As is shown in the table, only for the indicator livestock density does increasing premiums per hectare result in higher effect sizes. The Q-between test also rejects the null hypothesis of homogeneity amongst the average effect sizes of the three different groups. The Q-within statistic decreases slightly, but still indicates heterogeneity of the effect sizes within the groups. For the indicator N-fertilizer, the second group shows the largest effect size, and it is also one of the few cases where the Q-within test indicates homogeneity for that group. For the indicator grassland, the Q-between test now signifies heterogeneity among the average effect sizes between groups. However, the unexpected negative effect sizes remain in all the groups.

To summarize, in the second analysis the Q-between tests indicate heterogeneity, which means that the moderator '*average premium per hectare*' has a significant influence on the magnitude of the effect size. However, in the ideal case, additional to between-group heterogeneity, the Q-within tests should indicate homogeneity. This does not occur in this first moderator analysis. Unfortunately, the number of observations is not large enough for a more differentiated analysis.

II) Average farm size of participating farmers The results of the moderator analysis '*average farm size*' are presented in Table 3.4. As in the previous case, we performed two kinds of analyses, one with two groups and another one with three groups. In the first analysis, the first group contains all case-study areas where the average farm size of participating farmers is lower than 80 ha, and the second group all case-study areas where the average farm size of participating farmers is bigger than 80 ha. In this first analysis, the Q-between tests of all three indicators signify homogeneity between the effect sizes of the two groups. This means that this first analysis does not support the assumption that the moderator

Table 3.4 Results of moderator analysis 'average farm size'

2-groups Analysis	N-Fertiliser Hedges'g	Q	$P_{(Q)}$	Livestock Density Hedges' g	Q	$P_{(Q)}$	Grassland Hedges' g	Q	$P_{(Q)}$
< 80 ha	-1.54	21.85	0.00	-0.92	100.85	0.00	-0.87	103.71	0.00
> 80 ha	-1.59	30.36	0.00	-0.72	60.12	0.00	-0.81	66.07	0.00
Q between		0.03	**0.87**		0.84	**0.36**		0.063	**0.80**
Q within		52.21	0.00		160.97	0.00		169.78	
3-groups Analysis									
< 40 ha	-1.23	18.13	0.00	-1.92	24.35	0.00	-0.26	81.45	0.00
> 40 ha	-1.55	4.66	**0.1**	-0.16	45.89	0.00	-1.20	46.65	0.00
> 100 ha	-1.84	26.97	0.00	-0.89	54.47	0.00	-0.86	32.73	0.00
Q between		2.471	**0.29**		37.10	0.00		9.01	0.01
Q within		49.77	0.00		124.71	0.00		160.84	0.00

variable '*average farm size of participating farmers*' has a significant influence on the magnitude of the effect size.

In the second analysis, the first group contains those case study areas where the average farm size of participating farmers is smaller than 40 ha, the second group those where average farm size is between 40 and 100 ha, and the third group those with an average farm size of bigger than 100 ha. For the indicator N-fertilizer, the Q-between test still shows homogeneity of the average effect sizes of the three groups, indicating that even in this more differentiated analysis, the average farm size of participating farmers does not seem to be influential for the magnitude of the effect size of this indicator. For the other two indicators, the Q-between test shows heterogeneity between the average effect sizes of the three different groups. However, the Q-within test still indicates in all cases heterogeneity amongst the effect sizes inside the groups. Unfortunately, the number of observations is not large enough for a more differentiated analysis.

III) Average absolute value 1997 In this third moderator analysis, we divided the effect sizes of the different case-study areas into two groups. For the indicator N-fertilizer, the first group contains those case-study areas where the average absolute value in 1997 was lower than 40 kg/ha and the second group those where it was higher than 40 kg/ha. For the indicator livestock density, the first group comprises all case-study areas with less than 1.5 Livestock Units per hectare on average in 1997, and the second group those with more than 1.5 Livestock Units per hectare. For the indicator grassland, the two groups are characterized by less than, or more than 50 per cent grassland area per UAA in 1997. The results of the moderator analysis '*average absolute value in 1997*' are shown in Table 3.5.

The Q-between test signifies for the indicators N-fertilizer and livestock density heterogeneity between the average effect sizes of the two different groups. This implies that the average absolute value in 1997 seems to have a significant

Table 3.5 Results of moderator analysis 'absolute value 1997'

N-Fertilizer	Hedges' g	Q	$P_{(Q)}$	Livestock Density	Hedges' g	Q	$P_{(Q)}$	Grassland	Hedges' g	Q	$P_{(Q)}$
<40 kg/ha	-1.11	22.85	0.00	<1.5 LU/ha	-0.56	81.31	0.00	<50%	-0.73	75.73	0.00
>40 kg/ha	-1.93	22.28	0.00	>1.5 LU/ha	-1.10	74.64	0.00	>50%	-0.92	93.48	0.00
Q-between		7.12	0.01			5.86	0.02			0.63	**0.43**
Q-within		45.12	0.00			155.95	0.00			169.21	0.00

influence on the magnitude of the average effect size. As was expected, the case-study areas with a higher absolute level of the indicator have a higher average effect size than those with a lower level of that indicator. This means that, in areas with a higher absolute value of the indicator in 1997, a higher percentage of the change rates of non-participating farmers lie below the average change rate of participating farmers. For the indicator grassland, the Q-between test reports homogeneity between the average effect sizes of the two groups. The Q-within tests show in all cases heterogeneity amongst the effect sizes. However, again, the number of observations is not large enough for a more differentiated analysis.

IV) and V) Intensive versus extensive farming, husbandry versus arable farming The results of the last two moderator analyses are given in Table 3.6. The table shows that the effect size for intensive farming is slightly lower than that of extensive farming, and that the effect size for arable farming is lower than that of mixed farming. However, the Q-between test signifies that the null hypothesis of between-group homogeneity cannot be rejected in both cases. This means that, according to this analysis, whether the case study area is characterized by intensive or extensive, or arable or husbandry, agricultural production structure does not have any influence on the magnitude of the effect size. The Q-within tests indicate, as in most of the previous moderator analyses, that there is still heterogeneity among the effect sizes in the two groups.

Table 3.6 Results of moderator analysis 'intensive-extensive' and 'arable-husbandry'

	N-Fertilizer						
	Hedges' g	Q	$P_{(Q)}$		Hedges' g	Q	$P_{(Q)}$
Intensive	-1.488	11.904	0.008	**Arable**	-1.477	43.349	0.000
Extensive	-1.666	40.000	0.000	**Husbandry**	-1.871	7.681	0.006
Q-between		0.336	**0.562**	Q between		1.208	**0.272**
Q-within		51.904	0.000	Q within		51.032	0.000

3.6 Conclusions and discussion

In this chapter, a first effort was made to apply the statistical methods of meta-analysis to the question of agri-environmental policy evaluation in the European Union. Because of the limited data availability, this study can rather be seen as exploratory and as a first test of how meta-analytical techniques handle the data. In spite of that, some general conclusions can be drawn on the basis of this analysis. First, the fact that meta-analysis artificially increases the sample size becomes visible in the results of Step 2, the combination of the effect sizes. This means that, although most of the original case studies show insignificant differences between the change rates of participating and non-participating farmers, the combined effect sizes demonstrate that there is an overall difference between the change rates. In other words, there is an indication that the agri-environmental policy intervention does indeed have a positive effect on the behaviour of participating farmers with respect to the chosen indicators. Furthermore, from the moderator analysis, it can be concluded that the variables '*average premium per hectare*' and '*average absolute value in 1997*' have a significant effect on the magnitude of the effect sizes, meaning that they influence the percentage level at which the change rate of non-participating farmers lies below the average change rate of participating farmers. In general, the effect sizes of the indicator N-fertilizer show the highest value. The reason for this could be that the reduction of N-fertilizer is easier to organize and less dependent on other conditions than the reduction of livestock density or the increase in grassland area. The number of livestock kept by a farmer is rather susceptible to current prices of meat and livestock, which might outstrip the payments from agri-environmental programmes. The effect sizes of the indicator share of grassland area per UAA all show unexpected negative signs. These paradoxical results may be because the indicator grassland area is a very broad measure, being subject to multiple decision-making processes, including some outside the agricultural sector as, for instance, in urban and landscape planning.

A prevailing problem throughout all moderator analyses is that the Q-within tests signify heterogeneity of the effect sizes within the different groups. The occurrence of this problem actually underlines the diversity of the European landscape and the differences in the structure of the agricultural sector, a fact which is often emphasized by researchers when they try to evaluate European agri-environmental policy. The methods of meta-analysis might be able to shed more light on this diversity if a larger enough number of observations were to become available. It would then be possible to apply more advanced methods of meta-analysis, such as multi-factor analysis, by taking into account two or more moderator variables or meta-regression analysis. With the limited number of observations available to us in this chapter, it is not possible to derive any more sophisticated conclusions from them. In the study they made for the European Commission, the researchers of the FAIR project recommended the introduction of monitoring programmes where-by the behaviour of participants and non-participants can be compared. With such a quasi-experimental impact assessment, it should be easier to compare policy outcomes with policy objectives. Quasi-experimental case study results would also increase the amount of potential input data for meta-analysis. Retrieving case studies using such an approach is certainly the most important task for improving and strengthening the meta-analysis as it is performed in this chapter.

References

Andersen, E., Primdahl, J., Oñate, J., Aguirre, J. Cummings, C., Peco, B., Schramek, J., Klingler, B. and Knickel, K. (1999), Analysing the Environmental Effects of Regulation 2078/92, in: D. Biehl, H. Buller and G. Wilson (eds), *Implementation and Effectiveness of Agri-Environmental Schemes Established under Regulation 2078/92.* Project FAIR 1 CT95-274, Final Consolidated Report, Volume II (annexes), Institut fuer laendliche Strukturforschung (Research Institute for Rural Development), Frankfurt.

Baaijens, S. and Nijkamp, P. (2000), 'Meta-analytic Methods for Comparative and Exploratory Policy Research', *Journal of Policy Modeling*, vol. 22, no. 7, pp. 821–858

Beinat, E. and Nijkamp, P. (1998), 'Environmental Rehabilitation: Efficiency and Effectiveness in Soil Remediation', *Studies in Regional and Urban Planning*, vol. 6, pp. 83–101.

Beinat, E. and Nijkamp, P. (eds) (1999), *Multicriteria Analysis for Land-Use Management*, Dordrecht, Kluwer.

Bergh, J.C.J.M. van den (1996), *Ecological Economics and Sustainable Development*, Cheltenham, UK, Edward Elgar.

Bergh, J.C.J.M. van den and Button, K.J. (1997), 'Meta-analysis of Environmental Issues in Regional, Urban and Transport Economics', *Urban Studies*, vol. 34, nos. 5–6, pp. 927–944.

Bergh, J.C.-J.M. van den, Button, K.J., Nijkamp, P. and Pepping, G. (1997), *Meta-analysis in Environmental Economics*, Dordrecht, Kluwer.

Biehl, D. (1999), Agri-Environmental Policies in a Wider Theoretical and Institutional Context, in: J. Schramek, D. Biehl, H. Buller and G. Wilson (eds), *Implementation and Effectiveness of Agri-Environmental Schemes Established under Regulation 2078/92*. Project FAIR 1 CT95-274, Final Consolidated Report, Volume I (main report), Institut fuer laendliche Strukturforschung (Research Institute for Rural Development), Frankfurt.

Brouwer, F.M. and Berkum, S. van (1996), *CAP and the Environment in the European Union*. LEI-DLO, Wageningen, Wageningen Pers.

Brouwer, F.M. and Lowe, P. (1998), CAP Reform and the Environment, in: F.M. Brouwer and P. Lowe (eds), *CAP and the Rural Environment in Transition.* Wageningen, Wageningen Pers.

Buller, H. (2000), The Agri-Environmental Measure 2078/92, in: F. Brouwer and P. Lowe (eds), *CAP Regimes and the European Countryside: Prospects for Integration between Agriculture, Regional and Environmental Policies*, Oxford, Oxford University Press, pp. 199–220.

CEC (1992), *Council Regulation 2078/92 on the Introduction and Maintenance of Agricultural Production Methods Compatible with the Requirements of the Preservation of the Environment and the Management of the Countryside*, Luxemburg, Office for Official Publications of the European Community.

Chomitz, K.M. and Gray, D.A. (1996), 'Roads, Land Use, and Deforestation: A Spatial Model Applied to Belize', *World Bank Economic Review*, vol. 10, no. 3, pp. 487–512.

Cooper, H. and Hedges, L.V. (1994), *The Handbook of Research Synthesis*, New York Russell Sage Foundation.

Deblitz, C. (1999), *Vergleichende Analyse der Ausgestaltung und Inanspruchnahme der Umweltprogramme zur Umsetzung der VO(EWG) 2078/92 in ausgewählten MitgliEdstaaten der EU*, Landbauforschung (agricultural research institute)Völkenrode, Sonderheft (special issues) 195, Braunschweig.

European Communities (1995–2000), at http://www.europa.eu.int/comm/regional_policy/activity/erdf/er1b1_en.htm (2000-03-09).

Finco, A. and Nijkamp, P. (2000), 'Towards a Sustainable Future of Cities in Europe' in: J. Stillwell and H. Scholten, (eds.), *Land Use Simulation for Europe*, Kluwer, Dordrecht, 2001, pp. 173–192.

Frederick, K.D. and Rosenberg, N.J. (1994), *Assessing the Impact of Climate Changes on Natural Resource Systems*, Kluwer, Dordrecht.

Giaoutzi, M. and Nijkamp, P. (1994), *Decision Support Models for Regional Sustainable Development*, Aldershot, UK, Ashgate.

Glass, G., MacGraw, B. and Smith, M. (1984) *Meta-analysis in Social Research*, Sage, Beverly Hills.

Groeneveld, R.A. and Ierland, E.C. van (2000), *Economic Modelling Approaches to Land Use and Cover Change*, Report no. 410 200 045, Department of Economics and Management, Wageningen University, Wageningen University.

Hedges, L.V. and Olkin, I. (1985), *Statistical Methods for Meta-Analysis*, New York, Academic Press.

Hunter, J.E. and Schmidt, F.L. (1990), *Methods of Meta-Analysis*. Sage Publications, Newbury Park.

Ijskes, E. (1999), 'A Meta-analytic Examination of the Antecedents of Budgetary Slack', Free University Amsterdam, Department of Accountancy.

Lier, H.N. van, Jaarsma, C.F., Jurgens, C.R. and Buck, A.J. de (1994), *Sustainable Land Use Planning*, Amsterdam, Elsevier.

Lowe, P. and Baldock, D. (2000), 'Integration of Environmental Objectives into Agricultural Policymaking', in: F. Brouwer and P. Lowe (eds), *CAP Regime and the European Countryside: Prospects for Integration between Agriculture, Regional and Environmental Policies*.

Matarazzo, B. and Nijkamp, P. (1997), 'Methodological Complexity in the Use of Meta-analysis for Empirical Environmental Case Studies', *International Journal of Social Economics*, vol. 34, no. 719, pp. 799–811.

Miller, J.M. and A.J. Plantinga, A.J. (1999), Modeling Land Use with Aggregate Data. *American Journal of Agricultural Economics*, 81, pp. 180–194.

Nijkamp, P. (2000), 'Critical Success Factors for Soil Remediation Policy', *Journal for Environmental Policy and Law*, no. 1, pp. 81–98

Nijkamp, P. and Vreeker, R. (2000), 'Sustainability Assessment of Development Scenarios: Methodology and Application to Thailand', *Ecological Economics*, vol. 33, no. 1, pp. 7–27.

OECD (1997), *Environmental Indicators for Agriculture*, Paris, OECD.

Parks, P.J. and Schorr, J.P. (1997), 'Sustaining Open Space Benefits in the Northeast: An Evaluation of the Conservation Program', *Journal of Environmental Economics and Management*, vol. 32, pp. 85–94.

Pearce, D. (1991), 'The role of carbon taxes in adjusting to global warming', *Economic Journal*, vol. 101, n° 407, pp. 938–948.

Petitti, D.B. (1994), *Meta-analysis, Decision Analysis and Cost-Effective Analysis for Quantitative Synthesis in Medicine*, New York, Oxford University Press.

Rosenberg, M.S., Adams, D.C. and Gurevitch, J. (1997), *MetaWin* (statistical software for meta-analysis with resampling test), Sunderland, Massachusetts, Sinauer Associates.

Rosenthal, R. (1991), *Meta-Analytical Procedures for Social Research*, Newbury Park, Sage Publications.

Rosenthal, R. (1994), 'Parametric Measures of Effect Sizes', in: H. Cooper and L.V. Hedges, (eds), *The Handbook of Research Synthesis*, New York, Russel Sage Foundation.

Schramek, J., Biehl, D., Buller, H. and Wilson, G. (1999), *Implementation and Effectiveness of Agri-Environmental Schemes Established under Regulation 2078/92.* Project FAIR1 CT95-274, Final Consolidated Report, Volume 1, Institut fuer laendliche Strukturforschung (Research Institute for Rural Development), Frankfurt.

Shaddish, W.R. and Haddock, C.K. (1994), 'Combining Estimated of Effect Sizes', in: H. Cooper and L.V. Hedges (eds), *The Handbook of Research Synthesis*, Russell Sage Foundation, New York.

Silvis, H.J. and Rijswick, C.W.J. van (1999), *Tussen Interventie en Vrijhandel, WTO en de Nederlandse Agrosector* (Between intervention and free trade, the WTO and the Dutch agricultural sector), Report 5.99.05, The Hague, Agricultural Economics Research Institute (LEI).

Soufi, R. and Tuddenham, M. (1995), The Reform of the European Union Common Agricultural Policy, in: R. Gale, S. Barg and A. Gilles (eds), *Green Budget Reform*, London, Earthscan Publications Ltd.

Chapter 4

A Comparative Analysis of Rurality at the EU Level and Turkey

Aliye Ahu Gülümser, Tüzin Baycan-Levent and Peter Nijkamp

4.1 Defining rurality

Civilization started with the Agricultural Revolution and has proceeded with industrialization. The beginning of settlements forms the roots of today's urban areas in which most of the world's population resides. However, this trend never undermined the importance of rural areas. The term 'rural area' was used originally as the home of agricultural activities in order to identify the areas which were not urban. However, defining rural areas as the non-urban space or the domain where agriculture and physical landscape are important is inadequate to describe today's complex reality (Labrianidis, 2006). The division of rural and urban areas becomes increasingly fuzzy. Consequently, rural areas are increasingly a part of the modern leisure industry, with mass tourism, on the one hand, and small-scale recreation on the other (Vaz et al., 2006). In other words, beyond the traditional rural productive function, viz. supplying agricultural, agro-food, forestry goods, goods from extractive industries and craft products, rural areas have become an environment for living and leisure activities (Léon, 2005).

'Rural area' is an often used term in policy circles, as well as in the scientific community and public debates; nevertheless, there is no unequivocal definition of this term, which often combines regions with many diverse features (Baum et al., 2004). Rural is a fuzzy concept which is contested in terms of identifying the critical parameters of rural space (Halfacree, 1993; Pierce, 1996). Various ways of classification and different definitions in the literature are derived to define 'rurality', including the level of population density, the rate of population loss or gain, settlement size, local economic structure and landscape (Akder, 2003; Ballas et al., 2003; Baum et al., 2004; Bryden, 2002; Ilbery, 1998; Labrianidis, 2004). Actually, the meaning of rurality depends on the perception of each individual who integrates visions of rurality into everyday life (Ilbery, 1998; Hoggart et al. 1995; Halfacree, 1995). Moreover, rural has also been used in different contexts from developed countries to underdeveloped ones (Dinis, 2006). The developmental processes of social, economic and political restructuring in many countries are reshaping rural areas (Woods, 2005; Labrianidis, 2006), and this pushes governments to focus more on them.

Especially in the EU the future of rural peripheries, as well as the future of rural societies, is becoming an important development and planning issue. The globalization, liberalization, free market activities and changes of cultural values have led rural areas to become more consumerized and more externally interrelated (Labrianidis, 2006). 80 per cent of Europe is now rural, sheltering 25 per cent of its population (van Leeuwen, 2006). The European Commission describes rural areas as complex economic, natural and cultural locations, which cannot be characterized by a 1-dimensional criterion such as population density, agriculture or natural resources (European Commission, 1999:23). Rural areas considered in terms of their cultural, social, political, and economic aspects – and particularly in terms of their futures – have attracted considerable attention from governments. From this perspective, the rurality of a candidate country is often the last negotiation issue taken into consideration by the EU. Turkey as the most discussed candidate is now in the accession period and during the negotiations its rurality will certainly be addressed. The complexity of Turkey's rurality is recognized by the EU and the academic world. However, its rurality has not yet been evaluated as a whole.

Against this background, the aim of this chapter is to compare and evaluate Turkey's rurality with that of the EU countries on the basis of selected rural indicators. The data and information used for the comparison and evaluation of 26 countries are based on Eurostat and World Bank data. A multidimensional classification technique, factor analysis, is deployed to reduce 15 indicators, while 5 main factors, viz. underdevelopment, demography, urbanization, higher education and industrialization levels are used to define Turkey's rurality in the European context. In the next section, a literature review is provided in order to identify rural indicators that are often used to measure the rurality of a region. In the following section, the EU's and Turkey's rurality will be compared while giving information about the data and methodology of the study. The chapter will conclude by discussing the results of the study and then proposing some guidelines for further studies.

4.2 Rural indicators and classifications

The classification of rural areas and the distinction between rural and urban areas are not easy tasks. Each country has its own definition usually focused on socio-economic indicators, and these are not globally applicable (Politechnico di Milano, 1999). However, in the global context, two main perspectives of rural typologies have been developed by the OECD and the EU.

The OECD (1994, 1996, 2003), in creating territorial and rural indicators, aims to be able to compare sub-national territories. According to the OECD, territorial studies have four main indicators: population and migration; economic structure and performance; social well-being and equity; and environment and sustainability (Table 4.1). The OECD definition of rural areas distinguishes two hierarchical levels of territorial unit, viz. local and regional. At the local

Table 4.1 Basic set of indicators and sub-criteria of OECD

Population and Migration	Social well-being and equity
Density	Income
Change	Housing
Structure	Education
Households	Health
Communities	Safety
Economic structure and performance	**Environment and sustainability**
Labour force	Topography and climate
Employment	Land use changes
Sectoral shares	Habitats and species
Productivity	Soils and water
Investment	Air quality

Source: Akder, 2003.

community level (administrative or statistical units – equivalent to NUTS5), the OECD identifies rural areas as communities with a population density below 150 inhabitants per square kilometre. At the regional level (aggregated sub-national regions – equivalent to NUTS3), the OECD distinguishes larger functional or administrative units by their degree of rurality, depending on what share of the region's population lives in rural communities. To facilitate the analysis, regions are clustered into three types:

1. Predominantly Rural Regions: with over 50 per cent of the population living in rural communities;
2. Significantly Rural Regions: with 15 to 50 per cent of the population living in rural communities;
3. Predominantly Urban Regions: with less than 15 per cent of the population living in rural communities.

On the other hand, the EU's rural typology is less strict and is changing over time as well. Eurostat, focusing on the degree of urbanization as a main indicator, developed an approach to define zones at the NUTS5 level. In this approach, EU regions are classified into 3 types:

- Densely Populated Zones: these are groups of contiguous municipalities, each with a population density above 500 inhabitants/km^2, and a total population for the zone of at least 50,000 inhabitants.
- Intermediate Zones: these are groups of municipalities, each with a density above 100 inhabitants/km^2, not belonging to a densely populated zone. The zone's total population must be at least 50,000 inhabitants, or it must be adjacent to a densely populated zone.

- Sparsely Populated Zones: these are groups of municipalities not classified as either densely populated or intermediate (Politecnico Di Milano, 1999; Ballas et al., 2003).

As a second EU rural typology, the classification of territories developed in the study programme of the European Spatial Programme can be shown. A specific typology of six broad types of territories is distinguished on the basis of urbanization rate; rural population density, the degree of contrast in the distribution of settlement size; average distance to any urban settlement; the primacy of the largest city; and the size of the largest centre at the NUTS 3 level (SPESP, 2000). This typology is as follows:

1. Regions dominated by a large metropolis,
2. Polycentric regions with high urban and rural densities,
3. Polycentric regions with high urban densities,
4. Rural areas under metropolitan influence,
5. Rural areas with networks of medium-sized and small towns,
6. Remote rural areas.

Besides these typologies of the EU and the OECD, there are also two different typologies which differ from the OECD and the EU typologies. One of those is developed by Politecnico di Milano and second is developed by three Greek researchers, Ballas, Labrianidis and Kalogeresis.

The Milan Approach is based on a strategic study of one new urban-rural partnership in Europe, and examines all European rural areas. The main assumption of this approach is that the diversity of rural areas and their heterogeneity is very great; it is impossible to develop a single and unequivocal definition of a rural area. In the study, an alternative methodology to describe the nature of rural areas based on the strengths and weaknesses of agricultural activities in Europe is identified. Their typology depends on the presence and absence in a specific area of four major indicators of a determined characteristic of rurality, viz. Productivity of Agriculture, Importance of Agriculture, Agricultural Compatible Activities, and Urban Sprawl (Table 4.2).

On the other hand, the Greek approach (called so here because the authors are Greek) attempts to draw a picture of European rural areas on the basis of a novel database, while comparing two different approaches: OECD and EUROSTAT. The aim of this approach is to create rural typologies on the basis of aggregative and disaggregative classification methods. They distinguish rural regions by means of four main indicators such as accessibility, dynamism-competitiveness, economic performance and the role of agriculture, and they exclude all urban regions from the analysis. As a result they reached a typology of 24 types of rural areas (see Table 4.3).

In addition to these general efforts to provide a classification of rural areas, there are also sectorally focused typologies within countries, e.g. policy-based

Table 4.2 Typology of rural areas: Milan approach

1	High productivity of agriculture	High importance of agriculture area			Strong
2		Low importance of agriculture area	High diversification of activities	Low urban sprawl	Strong
3				High urban sprawl	Under pressure
4			Low diversification of activities		Under pressure
5	Low productivity of agriculture	High importance of agriculture area	High diversification of activities	Low urban sprawl	Weak
6				High urban sprawl	Under pressure
7			Low diversification of activities		Weak
8		Low importance of agricultural area			Weak

Source: Politechnico di Milano, 1999.

Table 4.3 Typology of rural areas: Greek approach

	Accessibility	Economic performance	Dynamism	Importance of agriculture
1	Least accessible	Relatively low	Lagging	Dependent
2				Not dependent
3			Advancing	Dependent
4				Not dependent
5		Relatively high	Low competitiveness	Dependent
6				Not dependent
7			High competitiveness	Dependent
8				Not dependent
9	Semi-accessible	Low	Low competitiveness	Dependent
10				Not dependent
11			High competitiveness	Dependent
12				Not dependent
13		High	Low competitiveness	Dependent
14				Not dependent
15			High competitiveness	Dependent
16				Not dependent
17	Most accessible	Low	Low competitiveness	Dependent
18				Not dependent
19			High competitiveness	Dependent
20				Not dependent
21		High	Low competitiveness	Dependent
22				Not dependent
23			High competitiveness	Dependent
24				Not dependent
25	Urban			

Source: Labrianidis et al., 2003.

or as a tool for development plans or sectoral plans such as those for transport, education, health and housing etc. (Blunden et al., 1998; CIT, 2001; Cloke, 1977; Copus et al., 2001; Malinen, 1995; Reading et al., 1994; Satsangi et al., 2000; Williams et al., 2005).

The merits and generalization of the various typologies can be discussed from several perspectives. Here, we may find out the common indicators between these typologies. The distinction into rural and urban areas emerged as a result of policy issues or planning problems, e.g. to measure differences in the degree of rurality, etc. (Cloke, 1977; Scottish Executive Development Department, 2005). However the most important reason was to balance national and local perspectives. The criteria used in different typologies are endless concerning the diversity and uniqueness of rural areas.

4.3 A comparative analysis of rurality: Turkey in the EU-25

4.3.1 Prefatory remarks

Rurality normally attracts attention in terms of cultural, social, political, or economic aspects, and especially in terms of the future of rural areas. Rurality is considered by the EU as a combination of economic, natural and cultural components; it cannot be characterized by one-dimensional criteria such as population density, agriculture or natural resources. Therefore, early attempts to evaluate rurality were to measure it by its separate components, such as agriculture, demography or social well-being. On the other hand, the definition of rurality may change according to the perception of developed or developing countries. From this perspective, Turkey offers a complex picture as perceived by the EU and the academic world. However, a comprehensive evaluation of rurality was not carried out in this study. We took into consideration only the EU Member States and Turkey.

The present chapter investigates the rurality of the EU Member States, while comparing them with Turkey's rurality based on various selected data. The data and information used for the comparison and evaluation of 26 countries are based on Eurostat and World Bank data. A multidimensional classification technique, factor analysis, is applied to compare rurality within the EU Member States while reducing 15 indicators to 5 main factors, viz. underdevelopment, demography, urbanization, higher education and industrialization levels.

The 15 variables, for 26 countries at country level, used in the analysis can be seen in Table 4.4. While employing these 15 variables, the previous rural classifications were used and the intersection of these classifications was carried out. As can be seen from Table 4.4, variables are classified under the subtitles 'population', 'employment', 'income', 'education', 'land use' and 'environment-energy', mainly based on the OECD's classification. On the one hand, well-known variables to compare rurality such as population, land, population density, households or GDP are not included in the analysis as they are correlated with the rest of the variables. On the other hand, new variables related to the changing definition of rurality such as innovation, export and import rates also cannot be included in the analysis as the data obtained possessed missing values in terms of these variables. However, although some variables are not included in the analysis, the selection of the variables included in the analysis allows us to take into consideration different aspects of the socio-economic and demographic character of the countries, as early attempts to define and measure rurality are based on a geographical, socio-economic and demographic database. To compare countries in terms of their ruralities, we used shares and rates which enabled us to compare countries more reasonably by using the variables included in the analysis (Table 4.4).

The next sub-section will evaluate the rurality of the EU-25 and Turkey. The evaluation is based on 5 factors carried out by a factor-analytic approach which enables us to see the similarities and differences within the EU and between the EU and Turkey.

Table 4.4 Variables included in the analysis

Code	Description of the variables	From
POPULATION		
CBR	Crude Birth Rate	Eurostat
CDR	Crude Death Rate	Eurostat
PG	Population growth (annual %)	World Bank
EMPLOYMENT		
AES	Agricultural Employment per Total Employment	Eurostat
IES	Industrial Employment per Total Employment	Eurostat
SES	Services Employment per Total Employment	Eurostat
EDUCATION		
SE1	School enrolment, primary (% gross)	World Bank
SE2	School enrolment, secondary (% gross)	World Bank
SE3	School enrolment, tertiary (% gross)	World Bank
INCOME		
IOID	Inequality of income distribution	Eurostat
LAND USE		
ALS	Agricultural land per Total land area	Eurostat
LOM	Length of Motorways	Eurostat
NOD	Number of Dwellings	Eurostat
ENVIRONMENT – ENERGY		
CO^2	CO^2 emissions (metric tons per capita)	World Bank
EPC	Electric power consumption (kWh per capita)	World Bank

4.3.2 Rurality of European Union Member States: a factor-analytic approach

As mentioned above, there are many multivariate techniques that can be used to measure rurality. One of the well-known techniques is factor analysis which can be used to analyse interrelationships between a large number of variables and to explain these interrelationships in terms of their common underlying dimensions. Factor analysis is an interdependence technique in which all variables are considered, as each relates to all others, and where the concept of the variate, the linear composite of variables, is employed (Hair et al., 1998).

In this study, principal component analysis is used to transform the set of originally mutually correlated variables into a new set of independent variables. It is a non-stochastic approach, and it only deals with the common variance of the original variables. It first derives the first factor or the first principal component, which is supposed to account for the greatest part of the common variance. The second factor is supposed to account for the next greatest part of the common variance, and so on. A minimum part of the common variance is set, and factors below this critical level are eliminated. The relative lengths of the lines that express the different variable combinations are called eigenvalues. As the result of the principal component analysis is 82, this is 60 per cent of the variance of the original variables is explained (Table 4.5). A plot of eigenvalues of explained variances of factors is shown in the scree plot (see Figure 4.1). The first factor has an eigenvalue of 4.38 and the last factor's eigenvalue is above 1.

Table 4.5 Total variance explained

Component	Initial Eigenvalues			Extraction Sums of Squared Loadings			Rotation Sums of Squared Loadings		
	Total	Variance (%)	Cumulative (%)	Total	Variance (%)	Cumulative (%)	Total	Variance (%)	Cumulative (%)
1	4.38	29.19	29.19	4.38	29.19	29.19	3.64	24.26	24.26
2	3.34	22.24	51.43	3.34	22.24	51.43	2.94	19.59	43.86
3	1.96	13.09	64.52	1.96	13.09	64.52	1.97	13.16	57.02
4	1.59	10.60	75.13	1.59	10.60	75.13	1.93	12.85	69.87
5	1.12	7.47	82.60	1.12	7.47	82.60	1.91	12.72	82.60

From these findings, the factor analysis was carried out with 5 factors rotated with the equamax method. Loadings of the factors tend to be either high or low in absolute values (see Table 4.6). In the first component, the highly loaded variable is agricultural land per total land area but the other loadings are also high so that the component represents the underdevelopment level of the area (see Tables 4.6 and 4.7).

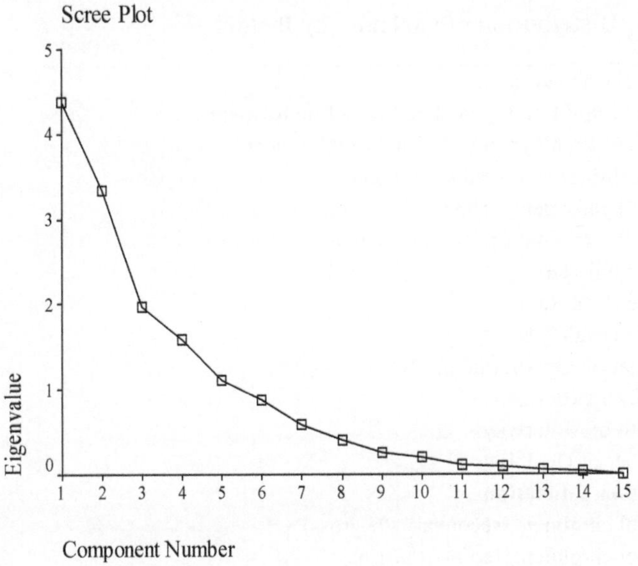

Figure 4.1 Scree plot of all factors

Table 4.6 Principal component matrix

Code[1]	Factor 1: Under-development	Factor 2: Demography	Factor 3: Urbanization	Factor 4: Higher Education	Factor 5: Industrialization
AES	0.84	-0.04	-0.18	-0.15	-0.24
SES	-0.78	0.00	0.08	0.36	0.38
IOID	0.81	0.16	0.12	-0.16	-0.19
SE1	-0.48	0.39	-0.04	0.33	-0.44
ALS	0.85	0.35	0.14	-0.16	0.00
CBR	0.57	0.69	0.00	0.10	0.18
CDR	-0.04	-0.89	-0.13	0.18	-0.11
PG	-0.01	0.96	0.06	0.00	0.11
LOM	-0.14	0.09	0.96	0.02	-0.03
NOD	0.15	0.02	0.97	0.07	0.03
SE2	-0.23	0.13	0.09	0.84	0.11
SE3	-0.02	-0.39	0.01	0.79	-0.25
CO2	-0.20	0.06	0.02	-0.27	0.78
EPC	-0.46	0.19	-0.09	0.25	0.64
IES	-0.21	-0.49	-0.03	-0.34	-0.57

Note: 1. For an explanation of the codes, see Table 4.7 below.

Table 4.7 Distribution of variables by factors

Factor 1: Underdevelopment	
AES	Agricultural Employment per Total Employment
SES	Services Employment per Total Employment
IOID	Inequality of income distribution
SE1	School enrolment, primary (% gross)
ALS	Agricultural land per Total land area
Factor 2: Demography	
CBR	Crude Birth Rate
CDR	Crude Death Rate
PG	Population growth (annual %)
Factor 3: Urbanization	
LOM	Length of Motorways
NOD	Number of Dwellings
Factor 4: Higher Education	
SE2	School enrolment, secondary (% gross)
SE3	School enrolment, tertiary (% gross)
Factor 5: Industrialization	
CO2	CO_2 emissions (metric tons per capita)
EPC	Electric power consumption (kWh per capita)
IES	Industrial Employment per Total Employment

The second component represents the demographic change of the area. Consecutively, the other factors represent the built-up area/urbanization level, the higher education level, and the industrialization level of countries.

The next issue is to analyse the commonalities of all variables to see the degree of reflection by the 5 main components. The highest commonalities are the ones related to built-up areas and population growth, while primary school enrolment has the lowest commonality. This shows us that school enrolment is not correlated with the other variables and its uniqueness is high.

The first factor measures the underdevelopment level of the countries by agricultural employment, inequality of income distribution, agricultural land, service employment and primary school enrolment. According to the underdevelopment factor, Turkey has the highest score, while Malta, Slovenia, Luxembourg, Austria, the Czech Republic and Sweden have the lowest scores (Table 4.8). Greece, Poland, Lithuania and Latvia come after Turkey, and they have a high level of relative underdevelopment (Figure 4.2).

As can be seen from Figure 4.2, most of the EU Member States have a negative score which means they are developed, but are quite different from each other. Therefore, the similarity within northern countries and also southern countries can be seen from Figure 4.2. However, Ireland and United Kingdom, which are

Table 4.8 Factor 1 – Underdevelopment

-0.95/-0.62	-0.61/-0.13	-0.12/0.32	0.33/1.13	1.14/4.61
Malta	Germany	Italy	Greece	Turkey
Slovenia	Netherlands	Ireland	Poland	
Luxembourg	Belgium	United Kingdom	Lithuania	
Czech Republic	Spain	Estonia	Latvia	
Austria	France			
Sweden	Cyprus			
	Slovakia			
	Portugal			
	Denmark			
	Finland			
	Hungary			

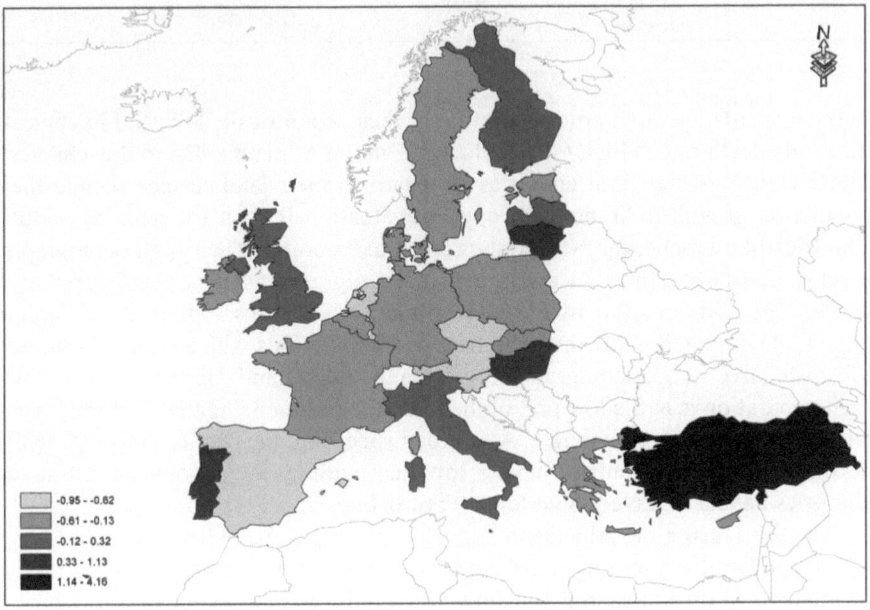

Figure 4.2 EU-25 and Turkey by underdevelopment level

northern countries, behave contrarily to the rest of the northern countries and have a higher value in terms of their underdevelopment level.

The second component, the demography factor, is used to measure demographic changes, with crude birth rate and population growth having a positive effect and crude death rates having a negative effect. According to this factor, Turkey and also Ireland have the highest scores; while Latvia, Lithuania, Estonia and Hungary have the lowest scores (Table 4.9). It is usually expected that developing countries have high crude birth rates and crude death rates together with high population growth, but, according to our results, new EU Member States like Latvia, Lithuania, Estonia, and Hungary have a low score.

Table 4.9 Factor 2 – Demography

-2.15/-1.41	-1.40/-0.32	-0.31/0.45	0.46/1.07	1.08/1.72
Latvia	Germany	Slovenia	France	Turkey
Lithuania	Poland	Denmark	Netherlands	Ireland
Estonia	Italy	United Kingdom	Luxembourg	
Hungary	Greece	Sweden	Malta	
	Czech Republic	Belgium	Spain	
	Finland	Austria	Portugal	
	Slovakia		Cyprus	

In other words, the high crude birth rate of these countries is low, and in contrast the crude death rate is high (see Table 4.A.1 in the Appendix 4.A to this chapter). Those countries are small countries in regard to their land surface so that their population growth is limited. An opposite situation is seen for more developed countries like France, the Netherlands and Luxembourg. Their high demography level is caused by their attractiveness to immigrants which affects population growth. In addition, the two islands Malta and Cyprus, which have limited attraction compared with other countries and populations with a relatively limited life span, have a high demography level. On the other hand, Germany which has a high population is in the low part of the rank, the reason being that Germany has a negative population growth rate. As a broad spectrum, the results show that while northern European countries have a low demography level, southern European countries have a relatively high level (Figure 4.3).

The third factor, the urbanization level is measured by the length of motorways which can be also a tool to understand the accessibility, but here we used it with the number of dwellings to define how much built-up land exists in the countries. From this point of view, Germany which is in the middle of Europe has the highest score (Figure 4.4). Germany, the United Kingdom, Italy, Spain and France are the countries which are far from the others in terms of their extreme urbanization score. The reason is that most of the European transportation projects are concentrated in those areas. Therefore, most of the peripheral countries like Ireland, Lithuania, Estonia, Latvia, and Finland and additionally Luxembourg, Slovenia, and Cyprus have a low urbanization level. On the other hand, Turkey is, on average, similar to high level countries, as it is strategically a bridge between Europe and Asia

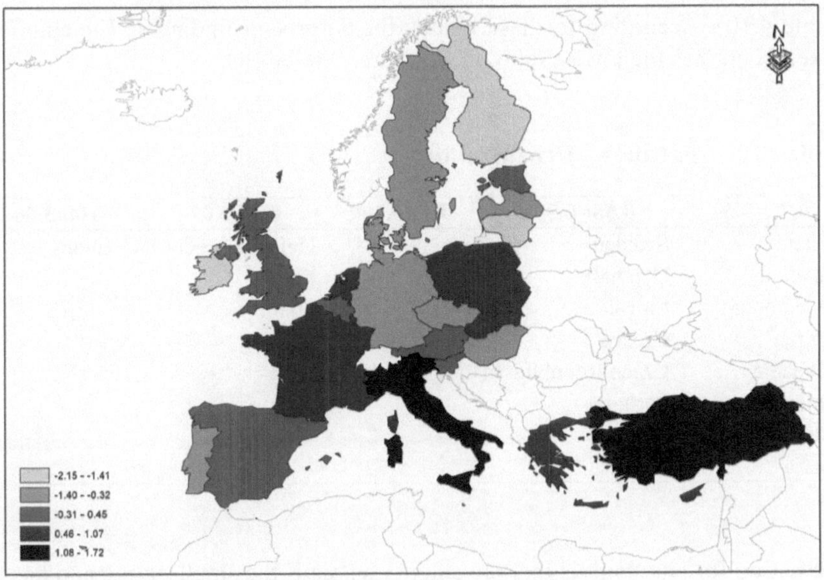

Figure 4.3 EU-25 and Turkey by demographic level
Note: Rate of natural population change.

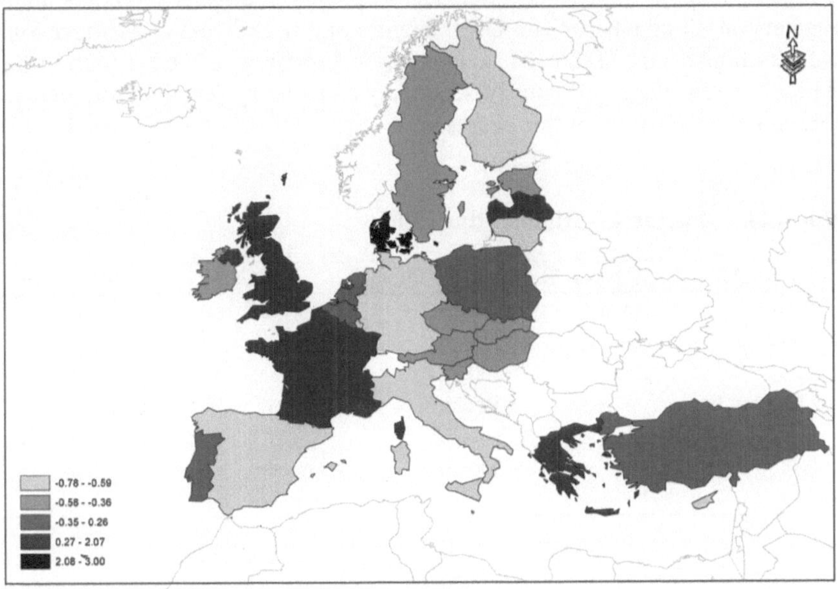

Figure 4.4 EU-25 and Turkey by urbanization level

(Table 4.10). According to these results the surprising finding is Luxembourg. Luxembourg has the lowest ratio of urban areas in the EU.

Table 4.10 Factor 3 – Urbanization

-0.78/-0.59	-0.58/-0.36	-0.35/0.26	0.27/2.07	2.08/3.00
Ireland	Sweden	Malta	United Kingdom	Germany
Slovenia	Denmark	Belgium	Italy	
Luxembourg	Slovakia	Poland	Spain	
Cyprus	Greece	Netherlands	France	
Lithuania	Czech Republic	Turkey		
Finland	Portugal			
Estonia	Hungary			
Latvia	Austria			

The reason for our result is that the country's area is the smallest one and has the lowest number of dwellings which depend on land area. The results show that southern and western European countries have a parallel tendency in terms of urbanization level and so do northern European countries.

Our fourth factor is the enrolment in secondary and tertiary schools which here are together called 'higher education', as secondary and tertiary schools are not obligatory in all countries. The UK, Belgium and Scandinavian Member States have very high scores (Table 4.11). This is not surprising, as these countries have the highest percentages in terms of secondary and tertiary school enrolments (see Table A in the Appendix of this chapter).

Table 4.11 Factor 4 – Higher education

-1.75/-1.27	-1.26/-0.78	-0.77/0.05	0.06/0.80	0.81/1.96
Luxembourg	Germany	Austria	Latvia	Denmark
Slovakia	Turkey	Hungary	Lithuania	Finland
Malta	Cyprus	Italy	France	Belgium
Czech Republic		Poland	Ireland	Sweden
		Estonia	Spain	United Kingdom
		Slovenia	Greece	
		Portugal	Netherlands	

On the other hand, Malta, Cyprus, Luxembourg, Slovakia, the Czech Republic and Turkey have a low higher education enrolment (Figure 4.5). The reason for Turkey's low score is the extreme remoteness of some areas, while at the same time

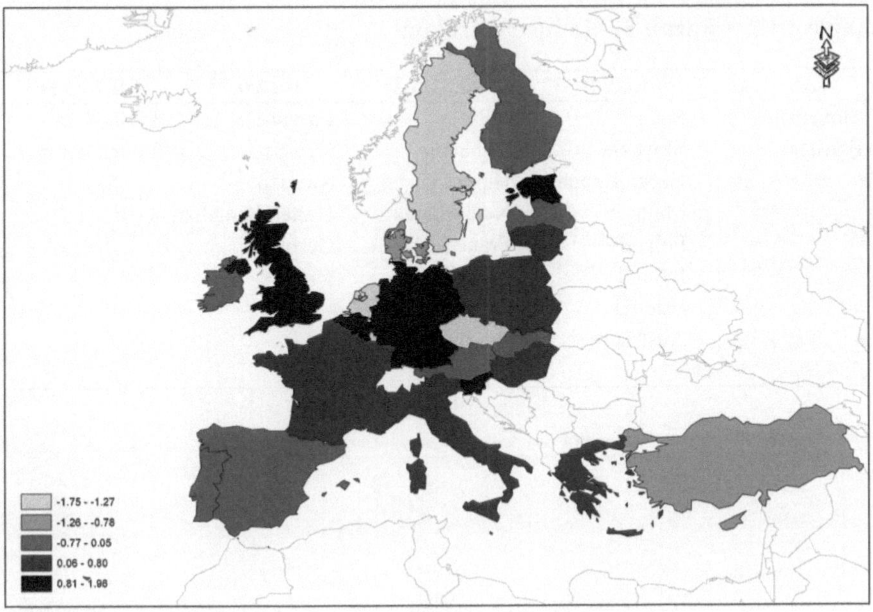

Figure 4.5 EU-25 and Turkey by higher education level

secondary and especially tertiary schools are not spread equally around Turkey. The unequal spread of schools does not only exist in Turkey but also, around Europe so that most of the European countries have a different share of school enrolment. Therefore, northern and western European countries have a similar level, and so do eastern and southern European countries.

The last factor is the industrialization level which evaluates three components, viz. the employment share of industry, CO_2 emissions, and the electric power consumption of the country. According to the results, Luxembourg and Finland have the highest scores and none of the other countries can match them. On the other hand, Slovenia is the least industrialized country. In terms of industrialization level, Turkey is in the middle which means that it has an industry which is not sufficiently developed. As can be seen in Figure 4.6, northern European countries are close in terms of having a high industrialization level; therefore, eastern and southern European countries are close in terms of their low industrialization level. It is a well-known reality that technology and innovation used in northern European countries is highly developed compared with southern European countries where the economy is more concentrated in the service sector, especially tourism.

On the basis of these results, we aim to calculate an overall rurality score from all the factor scores. When calculating this rurality score, underdevelopment and demography scores had a positive effect, and urbanization, higher education and industrialization had a negative effect. It is assumed that rural areas have fewer

Table 4.12 Factor 5 – Industrialization

-1.69/-1.41	-1.40/-0.33	-0.32/0.21	0.22/0.73	0.74/3.56
Slovenia	Spain	Italy	Latvia	Finland
Portugal	Slovakia	Lithuania	Cyprus	Luxembourg
	Czech Republic	Hungary	Sweden	
	Malta	Netherlands	United Kingdom	
	Poland	France	Germany	
	Ireland	Turkey		
	Estonia	Greece		
	Austria	Belgium		
		Denmark		

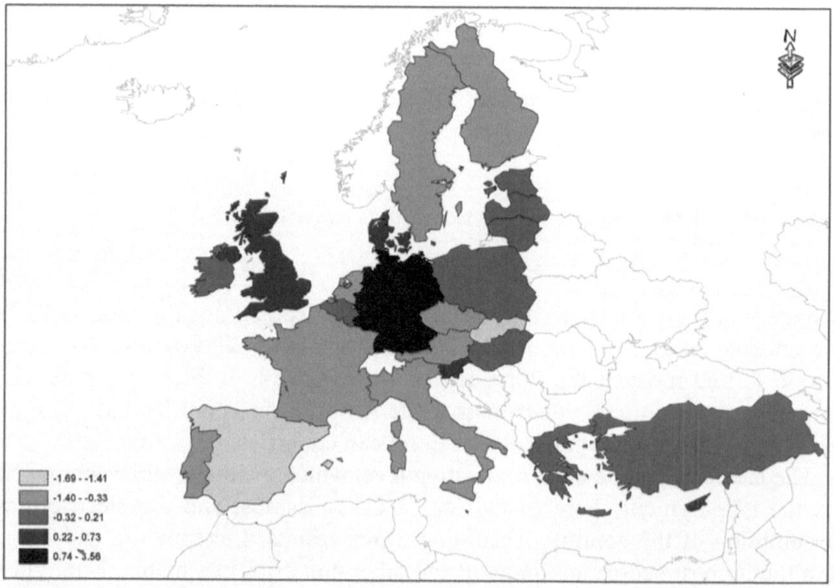

Figure 4.6 EU-25 and Turkey by industrialization level

dwellings and motorways, education facilities are often missing in these areas, and also those areas have not yet been industrialized.

In Europe, 11 countries have positive rurality levels, although they are rather different from each other (Table 4.13). According to the results of our study, the general picture is that northern countries and western European countries are not really rural; in contrast, southern and eastern European countries are rural (Figure 4.7). The UK and Germany have the lowest scores, although they both give importance to their rurality.

Table 4.13 Total factor – Rurality

-4.22/-2.81	-2.80/-1.45	-1.44/0.22	0.23/2.50	2.51/6.41
United Kingdom	Belgium	Greece	Ireland	Turkey
Germany	Italy	Estonia	Portugal	
Finland	France	Lithuania	Malta	
	Sweden	Netherlands	Slovakia	
		Spain	Cyprus	
		Hungary	Slovenia	
		Denmark	Czech Republic	
		Latvia	Austria	
		Luxembourg	Poland	

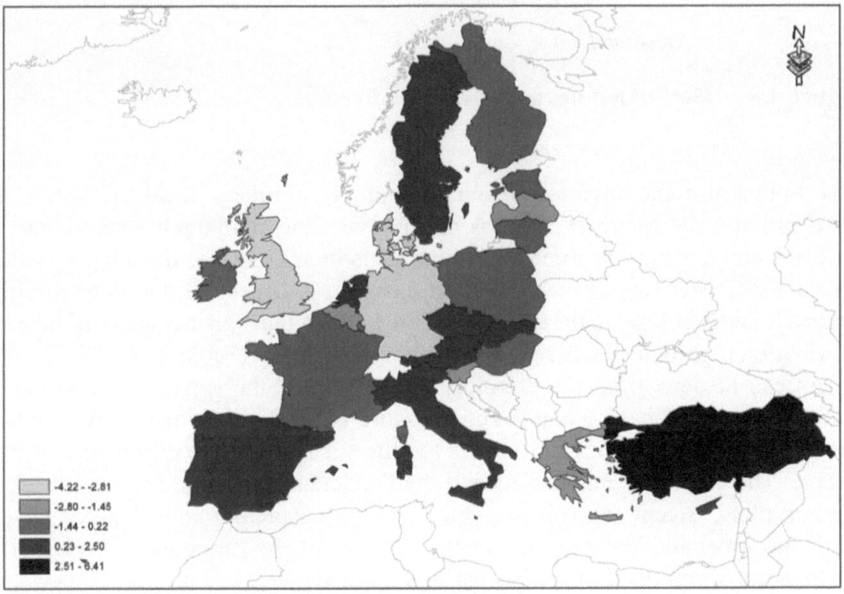

Figure 4.7 EU-25 and Turkey by rurality level

On the other hand, Turkey has the highest rurality level and none of the other countries is anywhere near its level (Table 4.14). Ireland follows Turkey in terms of being rural and has the second highest rurality score together with Portugal (Table 4.14). Rural Poland which is seen as similar to Turkey, has also a high score (0.82) but this is much lower than Turkey's. Rurality is obvious in the periphery of Europe (Figure 4.7).

The wide range of rurality scores can be better seen from a detrended q-q plot which is a tool to assess whether the shape of distribution is normal (Figure

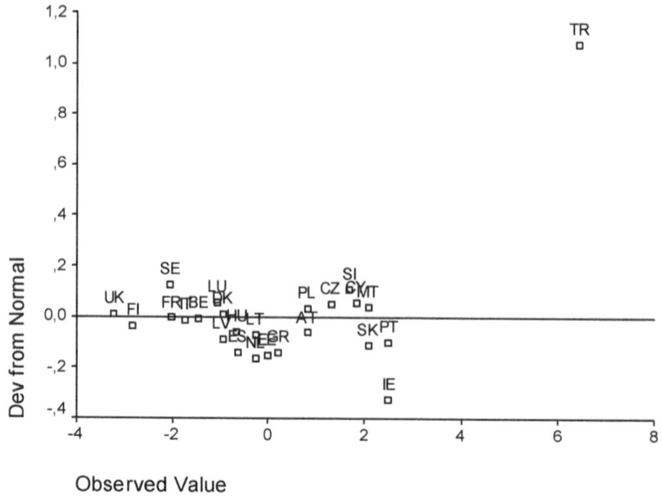

Figure 4.8 Detrended normal q-q plot of rurality

4.8). In this plot, the obviously different country is Turkey; however, without it, the distribution of Member States is not normal either. Although, over course of time, the convergence between EU countries is increasing, the divergence within the countries becomes greater. The best example is Greece. In terms of rurality, Greece is getting closer to the average for the EU, but the high inequality of income distribution highlights the divergence within the country (Table 4.14).

As can be seen from the boxplots of the factors, the spread of variables is quite different and has no equilibrium (Figure 4.9). Hence, it also shows us that each country, even if it has similarities to others, can also have differences as it is unique. The countries' uniqueness and their specialization can also be seen from the box plots, except for Turkey, which has high scores in almost every factor, while the upper and lower outliers in the Member States vary quite considerably.

In terms of rurality, Europe has various characteristics and different aspects, as stated in its rurality definition. A country which has a high level in one factor cannot be rural by having the lowest score from another factor. It is the consequence of the complexity and difficulty of measuring and defining rurality. For example, according to the results, Latvia has a high underdevelopment level but its rurality is less than zero (Table 4.14). The opposite can also be true, as Malta having the lowest underdevelopment level has a positive rurality score.

In other words, according to Factor 1, northern European countries have higher scores as agriculture is highly important and developed in those countries. However, the ones which have scores close to Turkey's are Latvia, Lithuania and Poland (which are new Members States) and Greece. On the other hand, in terms of Factor 2, the demographic level, southern and western countries are similar and have higher scores. In contrast, the new Member States, viz. Latvia, Lithuania and

Table 4.14 Factor scores by country

Country	F1	F2	F3	F4	F5	Rurality
Turkey	4.16	1.69	0.26	-0.85	0.04	**6.41**
Ireland	-0.03	1.72	-0.78	0.31	-0.35	**2.50**
Portugal	-0.21	0.95	-0.41	0.05	-1.41	**2.50**
Malta	-0.95	0.78	-0.26	-1.39	-0.61	**2.09**
Slovakia	-0.31	-0.32	-0.44	-1.51	-0.76	**2.09**
Cyprus	-0.33	1.07	-0.65	-0.78	0.34	**1.83**
Slovenia	-0.82	-0.02	-0.76	-0.08	-1.69	**1.69**
Czech Republic	-0.75	-0.38	-0.41	-1.27	-0.76	**1.32**
Austria	-0.71	0.45	-0.36	-0.40	-0.33	0.82
Poland	0.69	-0.75	-0.14	-0.24	-0.49	**0.82**
Greece	0.67	-0.41	-0.44	0.36	0.12	**0.22**
Estonia	0.32	-1.43	-0.59	-0.16	-0.34	**-0.02**
Lithuania	0.74	-1.58	-0.65	0.24	-0.17	**-0.26**
Netherlands	-0.42	0.73	-0.14	0.80	-0.08	**-0.26**
Spain	-0.35	0.87	1.61	0.36	-0.82	**-0.62**
Hungary	-0.13	-1.41	-0.37	-0.39	-0.13	**-0.65**
Denmark	-0.19	0.00	-0.48	1.02	0.21	**-0.94**
Latvia	1.13	-2.15	-0.59	0.21	0.30	**-0.94**
Luxembourg	-0.77	0.77	-0.75	-1.75	3.56	**-1.07**
Belgium	-0.37	0.31	-0.26	1.54	0.12	**-1.45**
Italy	-0.03	-0.66	1.59	-0.31	-0.22	**-1.74**
France	-0.34	0.67	2.07	0.30	-0.02	**-2.01**
Sweden	-0.62	0.31	-0.50	1.87	0.36	**-2.05**
Finland	-0.16	-0.32	-0.63	1.13	1.83	**-2.81**
United Kingdom	0.27	0.13	1.10	1.96	0.56	**-3.21**
Germany	-0.50	-1.04	3.00	-1.04	0.73	**-4.22**

Estonia, have far different scores than Turkey and have the lowest scores. In this sense, Turkey's score is close to that of many EU founding states, viz. France, the Netherlands and Luxembourg. Therefore, for the Factor 3, it is difficult to classify Member States spatially. However, western and southern European countries including Turkey are the most urbanized ones. For Factor 4 higher education enrolment, the distribution of states has a high variance, but similarities between northern and western and southern and eastern European countries can be seen. In this sense, Turkey is close to southern and eastern countries, where higher education enrolment is low. In terms of the industrialization factor there are again similarities between northern and western and southern and eastern European countries. In summary, it can be said that, apart from some exceptions, northern and western, on the one hand, and southern and eastern countries, on the other, are alike from many perspectives, and that Turkey is close to southern countries.

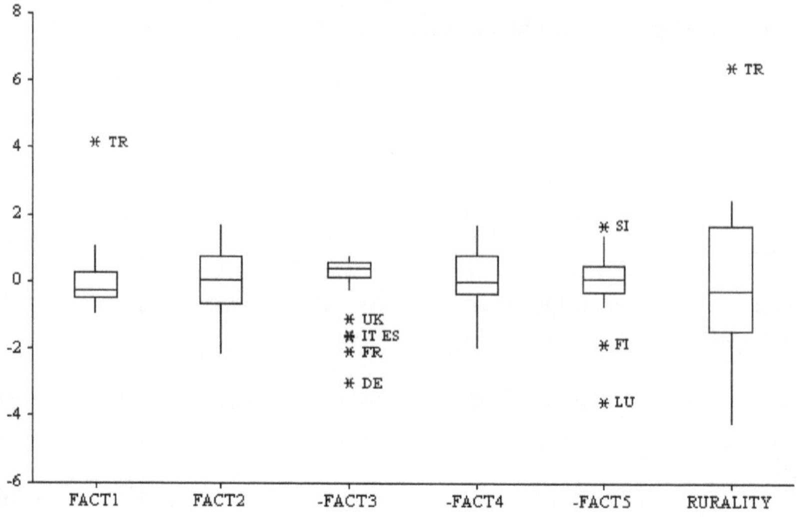

Figure 4.9 Boxplots

4.4 Concluding remarks

It is clear that rurality through its connections with agriculture has historic importance. Agriculture is the start of civilization and economies. So, even today, rurality is the main indicator of the socio-economic development of a country. Therefore, changing the definition of rural areas needs other components to differentiate the new rural areas from the traditional ones. Each country measures its rurality from a different perspective by their definition of 'rural'. Therefore, the EU is trying to introduce a common definition to evaluate each issue with a general and common understanding. In other words, the heterogeneity in the EU is an obvious and complex reality. The changing definition of rural and reformist views of the EU especially in its policies has brought to light how to compare non-member states with the Member States of the EU.

On the basis of early rural classifications, our aim was to draw a global picture of rurality by means of a holistic approach. Earlier studies have already emphasized the potential of Turkey in regard to agriculture, arable land, etc. However, this present study has given statistically an idea about how far Turkey is from the EU Member States on the basis of selected rural indicators. To look at the situation as a whole is difficult, however. Pairwises studies with Member States at the regional level would be more valuable to make a more precise comparison of Turkey's rurality.

The results of our study show that Turkey has the lower and upper values and even becomes sometimes an outlier in regard to each chosen indicator compared with the EU Member States. In other words, Turkey having the highest score of rurality is rather far from EU-25. The nearest countries to Turkey are Ireland and Portugal. Therefore, rurality of EU-25 is also different between all of its member states.

The reasons for Turkey's high rurality score are its high inequality of income distribution and the presence of less favoured areas. The presence of cities like Istanbul and Izmir and tourist places on the Mediterranean coasts of Turkey can hide the parts of Turkey which are not yet westernized. In other words, social and public facilities and other socio-economic facilities related to the quality of life are not equally spread around Turkey. This study emphasizes that the dynamics of Turkey's rurality depend more on socio-economic facilities that agricultural potential. Therefore, the authorities may focus on the elimination of the divergence which exists in Turkey, while developing relevant policies.

While evaluating this study, three questions come to mind: (1) Is Turkey so different from the EU Member States that this explains why it is still not a full Member? (2) Can rurality at the country level be as effective as at the regional level? (3) Do the variables used reflect rurality? Actually, the answer to the first question is still discussed in different fields, especially in politics and international studies (Buzan and Diez, 1999; Axt, 2005). The political view is that it is not the differences of Turkey but the political and strategic standing of the country that may affect Turkey's full membership of the EU. From the results of our study, in one sense this seems true, as even Member States have great differences between them.

Before answering questions 2 and 3, we should emphasize that to measure rurality based on any variables at any level will in any case be subjective and can only create a general framework to see the big picture from the components chosen by the researcher. In this context, for the second question, measuring rurality at the country level is not less efficient than measuring it at the regional level. However, to minimize the scale will be more effective so that relations and correlations will be more obvious, but more complicated. Rurality typologies in the literature are usually evaluated at the regional level, which is at the NUTS 3 and NUTS 5 levels, which offers in-depth analysis in terms of multi-dimensional evaluations about the regions. Therefore, the levels used to measure rurality, both country level and regional level, are effective in terms of what researchers want to reflect to their readers through their research. On the other hand, for the third question, we conclude that the representativeness of the variables used and the reflection of the level of rurality is obviously limited. Both the subjectivity in the selection of variables and the limitations in the number of variables will also lead to various restrictions in measuring and evaluating rurality. However, notwithstanding all the limitations inherent in our investigation, there is a need for more comprehensive research, so that the changing definition of rurality and development dynamics can be evaluated. This study is a first attempt for these kinds of studies and may offer a guideline for future investigations.

References

Akder, A.H. (2003), 'Linking Agricultural Statistics To Other Data Sources For Analysing Rural Indicators Of Social well-being and Equity', 8th IWG.AGRI Seminar, Perspectives for Agriculture and Rural Indicators and Sustainability, Château de la Muette, Paris, 21–22 November 2002 Statistics Directorate National Accounts – Agriculture, OECD.

Axt, H.J. (2005), Relations with Turkey and Their impact on the European Union. *Southeast European and Black Sea Studies,* vol. 5(3), pp. 365–378.

Ballas, D., Kalogeresis, T. and Labrianidis, L. (2003), 'A comparative study of typologies for rural areas in Europe'. Paper presented at the 43rd European Congress of the Regional Science Association, Finland.

Baum, S., Frohberg, K., Hartmann, M., Matthews, A. and Weingarten, P. (2004), *The future of rural areas in the CEE new member states*, Report of the Network of Independent Agricultural Experts in the CEE candidate countries, Germany.

Blunden, J. R., Pryce, W. T. R. and Dreyer, P. (1998), 'The classification of Rural Areas in the European Context: An Exploration of a Typology Using Neural Network Applications', *Regional Studies*, vol. 32, pp. 149–160

Bryden, J. (2002), *Rural Development Indicators and Diversity in the European Union.* Paper presented at the Conference on Measuring Rural Diversity, Washington, DC.

Buzan, B. and Diez, T. (1999), The European Union and Turkey, *Survival,* vol. 41 (1), pp. 41–57.

CIT (2002), Commission for Integrated Transport: Rural Transport: An Overview of Key Issues, at http://www.cfit.gov.uk/reports/rural/key/07.htm

Cloke, P. (1977), An index of rurality for England and Wales, *Regional Studies,* vol.11, pp. 31–46.

Copus, A., Petrie, S., Shucksmith, J., Shucksmith, M., Still, M. and Watt, J. (2001), *Pre-School Educational Provision in Rural Areas: A Report to the Scottish Executive Education Department,* Reproduced as an Arkleton Occasional Report No.1, Aberdeen.

Dinis, A. (2006), 'Rural Entrepreneurship: an innovation and marketing perspective', in: Teresa de Noronha Vaz, Eleanor J. Morgan and Peter Nijkamp (eds), *The New European Rurality: Strategies for Small Firms*, England, Ashgate.

European Commission, 1999, 'ESDP-European Spatial Development Perspective: Towards Balanced and Sustainable Development of the territory of the European Union', May.

Hair, J.F., Anderson, R.E., Tatham, R.L. Black, W.C. (1998), *Multivariate Data Analysis,* London, Prentice-Hall International.

Halfacree, K. (1993), 'Locality and social representation: space, discourse and alternative definitions of the rural', *Journal of Rural Studies,* vol. 9, pp. 23–37.

Halfacree, K.H. (1995), 'Talking about Rurality: Social Representations of the Rural as expressed by residents of six English Parishes', *Journal of Rural Studies,* vol. 11, pp. 1–19.

Hoggart, K., Buller, H. and Black, R. (1995), *Rural Europe: identity and change,* London, Edward Arnold.

Ilbery, B. (ed.) (1998), *The geography of rural change*, Essex, Longman.

Labrianidis, L. (ed). (2004), *The future of Europe's rural peripheries*. Ashgate Economic Geography Series, England, Ashgate.

Labrianidis, L. (2006), Human Capital as the critical factor for the development of Europe's rural peripheral areas, in: Teresa de Noronha Vaz, Eleanor J. Morgan and Peter Nijkamp (eds), *The New European Rurality: Strategies for Small Firms*, London, Ashgate.

Leeuwen, E.S. van and Nijkamp, P. (2006), 'The Embeddedness of Small Enterprises in the Rural Local Economy of Small and Medium-sized Towns', in: M.T. de Noronha Vaz, E. J. Morgan and P. Nijkamp (eds), *The New European Rurality: Strategies for Small Firms*, London, Ashgate.

Léon, Y. (2005), 'Rural Developmet in Europe: a research frontier for agricultural economists, *European Review of Agricultural Economists,* vol. 32 (3), pp. 301–317.

Malinen, P. (1995), 'Rural area typologies in Finland', Paper presented at the LEADER workshop 'Typology of European rural areas', 2–5 November, Luxembourg

OECD (1994), *Creating rural indicators for shaping territorial policy*, Paris, OECD.

OECD (1996), *Territorial Indicators of Employment; Focusing on Rural Development*, Paris, OECD.

OECD (2003), 'Territorial indicators of socio-economic patterns and dynamics' Working paper, 30 September.

Pierce, J. (1996), 'The conservation challenge in sustaining rural environments', *Journal of Rural studies,* vol. 12, pp. 215–219.

Politechnico di Milano (1999), A typology of rural areas in Europe, November, Milan.

Reading, R. Openshaw, S. and Jarvis, S. (1994). 'Are multidimensional Social Classifications of Areas Useful in UK Health-Service Research?', *Journal of Epidemiology and Community Health*, vol. 48, n° 2, pp. 192–200.

Satsangi, M., Bramley, G. and Storey, C. (2000), *Selling and Developing Land and Buildings for Rent and Low Cost Home Ownership: the Views of Landowners,* Edinburgh, Scottish Homes / Scottish Landowners Federation.

Scottish Executive Development Department (2005), *Rural planning typologies report*, February, Edinburgh.

SPESP (2000), *Final Report*, 31 March, Brussels, Stockholm.

Vaz, T. N, Morgan, J. E. and Nijkamp, P. (2006), *The new European rurality: strategies for small firms,* London, Ashgate.

Williams, N., Shucksmith, M., Edmond, H. and Gemmell, A. (1995), *Scottish Rural Life Update: A Revised Socio-Economic Profile of Rural Scotland*, The Scottish Office: Edinburgh.

Woods, M. (2005), *Rural Geography*, London, Sage.

Appendix 4A

Table 4.A.1 Data used in the analyses

Country	AES	SES	IOID	SE1	ALS	CBR	CDR	PG	LOM	NOD	SE2	SE3	CO2	EPC	IES
Austria	0.05	0.67	3.80	104.72	40.20	9.70	9.10	0.64	1670.00	3888821	100.35	48.71	1.07	8104.42	0.20
Belgium	0.02	0.73	4.00	104.90	45.70	11.10	9.80	0.43	1729.00	4745503	160.15	60.67	1.08	8411.94	0.18
Cyprus	0.05	0.71	4.10	97.61	14.70	11.20	7.00	1.01	268.00	323828	98.42	31.96	1.08	4758.63	0.13
Czech Republic	0.04	0.56	3.40	102.15	46.00	9.60	10.50	0.14	517.76	4344178	96.89	36.88	1.11	6070.08	0.30
Denmark	0.03	0.73	3.40	102.63	61.80	12.00	10.30	0.32	1010.00	2780658	127.31	66.83	1.08	6602.32	0.17
Estonia	0.06	0.59	5.90	100.37	17.00	10.40	13.20	-0.33	98.00	646764	95.91	64.49	1.11	5224.16	0.27
Finland	0.05	0.69	3.50	101.66	6.70	11.00	9.10	0.29	653.00	2624474	127.39	86.90	12.03	16426.83	0.19
France	0.04	0.00	4.20	105.21	53.70	12.70	8.40	0.59	10379.00	29600012	109.99	55.35	1.06	7816.45	0.00
Germany	0.02	0.66	4.40	99.39	47.70	8.60	9.90	-0.03	12037.00	38628607	100.07	50.10	10.03	6896.33	0.24
Greece	0.13	0.65	6.00	99.98	29.50	9.40	9.40	0.22	280.00	5709281	95.58	72.24	1.08	5040.51	0.14
Hungary	0.05	0.62	3.30	98.55	63.00	9.40	13.10	-0.22	542.00	4120551	103.41	51.89	1.05	3637.28	0.25
Ireland	0.06	0.66	5.00	105.57	61.30	15.20	6.90	1.08	176.00	1424565	109.02	55.29	1.10	6098.00	0.16
Italy	0.04	0.65	5.60	100.97	50.70	9.70	9.40	-0.13	6478.00	26681151	99.09	59.02	1.07	5619.84	0.22
Latvia	0.13	0.59	6.00	95.13	25.40	8.80	13.80	-0.54	0.00	1026168	94.70	70.98	1.02	2455.55	0.19
Lithuania	0.16	0.56	4.50	99.71	39.90	8.90	12.00	-0.54	417.00	1423463	102.52	68.99	1.03	3055.09	0.20
Luxembourg	0.02	0.78	3.70	99.18	49.50	12.00	7.90	0.74	115.00	217427	96.01	1.12	21.26	15935.21	0.10
Malta	0.02	0.68	4.60	102.59	30.40	9.70	7.20	0.57	2262.00	161658	93.93	29.92	1.07	4867.17	0.22
Netherlands	0.03	0.73	4.00	107.92	51.50	11.90	8.40	0.35	2289.00	6882317	121.94	58.00	1.09	6747.81	0.13
Poland	0.18	0.53	5.00	99.51	52.20	9.30	9.50	-0.04	405.00	12566780	104.51	59.47	1.07	3329.14	0.23
Portugal	0.12	0.57	7.20	118.48	41.50	10.40	9.70	0.58	1835.00	4152943	109.04	55.53	6.00	4383.18	0.20
Slovakia	0.05	0.56	5.80	100.25	45.60	10.00	9.60	0.05	312.80	1907196	91.73	33.99	1.06	5009.81	0.30
Slovenia	0.10	0.53	3.00	111.32	24.20	9.00	9.30	0.07	477.00	776965	111.80	70.12	1.07	6816.66	0.31
Spain	0.06	0.64	5.10	107.45	49.80	10.60	8.70	1.01	9739.00	21093061	116.52	63.55	1.07	5701.08	0.18
Sweden	0.02	0.75	3.30	109.11	7.70	11.20	10.10	0.40	1591.00	4336297	137.03	81.78	1.05	15402.63	0.17
Turkey	0.28	0.16	10.00	94.69	53.10	19.10	6.20	1.01	1775.00	17631782	85.30	28.01	1.02	1656.00	0.15
United Kingdom	0.01	0.76	5.30	100.82	67.20	12.00	9.70	0.48	3609.00	25957376	170.12	62.76	1.09	6209.24	0.15

Note: See Table 4.4 for an explanation of the abbreviations used for the variables employed in the analyses.

Chapter 5
Governance and the Determinants of Local Economic Development

Paulo Alexandre Neto, João Almeida Couto and Maria Manuela Natário

Introduction

Many authors (e.g. Lopes, 2001; Costa, 2002, 2003; Bramanti, 1999; Deryche, 2002; Guillaume, 1998; Braczyk et al., 1998; Cooke, 2003) have studied the mechanisms of governance, in particular the territorial process of innovation. Storper and Harrison (1991) draw attention to the structure of supply chains where the number of suppliers is different from the number of firms, in order to explain the asymmetry of power and to develop a typology of governance structures. In this case, governance is defined as a context in which different types of firms coexist and which generates different conditions for governance and interrelationships between companies.

According to Cooke (2003), the success of a company depends not only on the intra-organizational mechanisms of coordination and control but also on the structure of regional governance. Therefore, for Lopes (2001, p.150): 'The dynamics of territorial governance configure a virtual geographic space and promote synergies and competitiveness, through which complex relational processes linked to the integration of companies on their territory are associated with territorial articulation with the regional and national economy.' The governance concept applies to companies as well as to their economic environments. Governance includes the organizational forms and process through which economic activities in a specific field are coordinated and controlled.

The purpose of this chapter is to study the role of governance systems in the Raia Central Ibérica Region, which is located in the border area of Central Portugal/Spain, and present the factors that differentiate Portuguese Raia Central (RCP) and Spanish Raia Central (RCE) in terms of their local governance systems. The research assesses entrepreneur satisfaction with respect to the sub-regions' governance systems and their impact on the local dynamics of innovation and local economic development. We consider the results from an inquiry applied to the companies of the Raia Central Ibérica Region.

The study includes 169 companies, where agro-industries and the food chain represent about 21 per cent of total companies. We divided these companies into three groups using cluster analysis, and test the nature of satisfaction of the enterprises with governance systems, by considering local characteristics; their most important problems; the level of information and communication technologies (ICTs); the innovation and learning process; common identity, government/public intervention; and the cooperation relationships established.

We will examine differences in the satisfaction of enterprises with the governance systems in the sub-regions of Raia Central Ibérica. Based on an empirical study, the imbalances could be broadly attributed to: differences in problems; lack of common identity; lack of government/public intervention; factors such as industry, management education, and innovation; and lack of cooperation.

The chapter is structured as follows. First, we present a brief theoretical framework of governance and the methodology. Later, we evaluate the imbalances in the satisfaction of governance systems in the sub-regions of Raia Central Ibérica. Finally, we present some final reflections.

5.1 Literature review

The use of the governance concept regarding the regional economy dates from the early 1990s in the work of Storper and Harrison (1992) *Flexibilité, hiérarchie et dévelopment régional: les changements de structure des systèmes productifs industriels et leurs nuveaux modes de gouvernance dans les années 1990* published in the book of Benko and Lipietz *Les Regions qui Gagnent*. Storper and Harrison (1992, p.266) identify how territorial systems of industrial production function, and propose the governance concept, emphasizing the influence that company interrelationships have on the development and evolution process of these territorial systems. In regional analysis, the governance concept considers a set of existing or potential relationships between the companies in a local or regional territory that determines the collective model of functioning and economic development.

Governance has become a central topic among policy makers. There is an international consensus that policy making is evolving from a traditional top-down government approach towards a system of governing that focuses on engaging the citizens within an area (Cabus, 2003, p.2). Territorial governance is discussed extensively in the contemporary literature (Healey, 1997; Bogason, 2000; Castells, 2000; van Tatenhove and Leroy, 2000; Peters 2001; Le Galès, 2003; Brenner, 2004; Gualini, 2004; Pollit and Bouckaert, 2004; Salet, 2006; Voets and Rynck, 2006). Territorial governance, according to Domingues (1998, p.44), can be defined as being 'not only, the mere territorial government, but all the system of relations between institutions, organizations and individuals, which assures the collective choices and their accomplishment'. In this sense, the way in which different actors interact, the level of sophistication of these relationships, and their continuity and productivity are topics of growing interest for research.

Governance refers to rules, processes, and behaviour that affect the way in which power is exercised, particularly openness, participation, accountability, effectiveness, and coherence (European Commission, 2001, p.8). 'Improving governance – the way society collectively solves its problems and meets its needs

– is at the core of the government strategies to reconcile economic prosperity, social cohesion and environmental progress'. (OECD, 2001a, p.13).

The quality of territorial governance greatly depends on the territorial capacity to mobilize cooperation through public-public or private-private partnerships, yielding relational portfolios (Neto, 1999, p.929), and constituting a decisive aspect to the construction of developing collective territorial strategies. 'In a framework of good governance, government services across administrative levels co-ordinate their activities in order to enhance the global effectiveness of policies and minimize conflicting action. Civil society and the corporate sector are invited to participate in collective decisions and are encouraged to translate their involvement into concrete initiatives'(OECD, 2001a, p.13). The opportunities offered by the participatory approach to local and regional development are numerous (OECD, 2001a, p.14): i) a number of current concerns of the population at large are seen to have a better chance of finding an adequate policy response; ii) sustained economic growth in many countries allows more room to tackle geographical disparities; iii) partnerships may help to identify areas of synergy among activities based on local knowledge and skills, thereby allowing new economic development opportunities to be seized. The partnership concept was identified some 25 years ago as a way of maximizing mobilization and resources; it has since become a recurring feature of local development initiatives.[1]

A study by the OECD (1996) of local and regional partnerships in Ireland further characterizes the role of partnerships in improving governance frameworks. This study shows that partnerships provide models for broader participation in changing economies and societies. In 1999, the OECD published three other important studies on governance issues concerning best practices in local development, methodological contributions for local partnerships, and managing accountability in intergovernmental partnerships (OECD, 1999 a,b,c). In 2001, the OECD published two major studies about governance experiences in 'Local Partnerships for better Governance' and 'Governance in the 21st Century'. Other important studies on governance are *New Forms of Governance for Economic Development* (OECD, 2004), *Building Competitive Regions – Strategies and Governance* (OECD, 2005), *Uses and Abuses of Governance Indicators* (OECD, 2006) and *The New Rural Paradigm – Policies and Governance* (OECD, 2006).

Multilevel governance has also become a major issue in territorial policy making. New forms of governance targeting regional competitiveness are increasingly

1 For example, see the 1993 OECD survey on the partnership experiences in OECD countries as a response to increasing unemployment; the 1998 OECD survey on local and regional dimensions of partnerships around labour market policies, which demonstrated that partnerships involving key actors may lead to greater consistency between resource allocation and local priorities; and the 2000 and 2001 OECD studies on urban renaissance, local partnerships, and metropolitan governance (OECD, 2000 a,b; OECD, 2001 a,b,c).

oriented to vertical[2] and horizontal[3] cooperation. Firms, especially SMEs, are dependent on the environment in which they are located, which provides them with different types of 'local collective competition goods'[4] (OECD, 2005, p.12). 'The endurance and sustainability of the processes of territorial development lay on the collective capacity of mobilization, organization, and valuation of resources by the local actors' (Fermisson, 2005, p.2).

Inter-governmental relations and public policy making should become increasingly negotiated and contextual, stimulating the development of new state spaces and pressuring old scales (Voets and Rynck, 2006, p.906). Old governance relationships in local and regional territories are defined as rather fixed, formal-legalistic arrangements, while new governance spaces are more provisional, functional, collaborative, process-oriented patterns with loosely coupled actors from diverse institutional backgrounds and dispersed power (Hendricks, 2006). The local/regional development coalitions must be positioned within existing relationships; they have the potential to be effective if they commence on the assumption that the rules of the game are defined more by the market than by supposed interdependences (Cabus, 2003, p.5). Naturally, the capacity of mobilization in a specific territory is limited by the institutional capabilities that characterize it, and the number of the relevant actors and agents available.

Regional competitiveness policies and strategies pose significant governance challenges (OECD, 2005, p.8): i) regional competitiveness depends on the effective integration of sectoral policies such as R&D and education; ii) they also demand cooperation across levels of government and neighbouring regions, not to mention resource-sharing; iii) they also involve a high degree of coordination in planning and spatial development; iv) as important as the choice of strategy is the framework in which the strategy is implemented. Thus, governance is naturally associated with the recognition of traditional political-administrative insufficiencies, as well as the need for the State to have a transformative role and a wider vision of government (Le Galès, 2003, p.35).

One of the essential elements in the analysis of territorial governance conditions is the evaluation of the different actors in terms of their capacity and intervention in the territory's political-administrative action. To a great extent, the territory's government authorities are those which have a particular responsibility for the creation of territorial conditions at the political level and are responsible for the infrastructure and equipment. Their way of relating with the remaining authorities, companies, and institutions determines their own territorial governance quality. Therefore, their evaluation of political-administrative authorities is particularly

2 Cooperation between lower and high levels of government or cooperation between firms along the production chain.

3 Cooperation between firms or cooperation between different municipalities.

4 Crouch et al. (2001). Local collective competition goods include: availability of relevant skills; access to information related to technical evolution or external markets; and the sharing of a territorial label.

important for developing collective territorial strategies that are effectively mobilizing.

The local and regional territories are spaces of collective action; the results of their action depend on accurate knowledge and identification of the behaviour and relationships of the most relevant territorial agents, as well as their ability to coordinate. The success of territorial governance results from the procedures and mechanisms used to mobilize different institutional and managerial abilities. The possibility to mobilize these abilities strongly depends on the conditions for developing new ways of interaction that may reconcile the individual strategies of the multiple actors into legitimate joint strategies.

Organizer support, strategic management, and the recognition of regulation forms are crucial factors for the trajectory of these territories' development (Fermisson, 2005, pp. 43–45). Governance is a form of making decisions, obtaining consensus, and identifying strategies for collective action. The creation of institutional mechanisms for governance is particularly necessary given the natural existence of a rationalizing logic, automatic and integrative of the territory actions; the need to assure the leadership of the local and regional processes of development; the need to assure the conditions of collective appropriation of the territorial strategies; the possibility of guaranteeing the sophistication of the territorial strategies; the need for participative implementation of development strategies; the relevance of assuring the development of different institutions; the need to formalize collective strategies of territorial strategic planning.

Governance is not necessarily pacific; it is a local/regional source of tension based on unequal relationships between actors due to distinct rationalities. Therefore, it is important to create territorial conditions for the regulation of governance to be productive and focused. For this purpose, it is necessary to conceive and build an institutional form of support for governance: the territorial system. The institutionalization of relationship models of territorial systems of guidance is a crucial aspect for success. 'The regional and local development is, sometimes, more an approach of rationalities rather than a rationalized process' (Guerra, 2000, p. 52).

The question of collective leadership is also a fundamental issue for governance along with the question of creating new forms of engagement between public and private actors that assures continuity and direction to their own collective action. 'The clarification of engagements and the set-up of the action process and collective regulation is a slow game that is going to be built by action' (Guerra, 2000, p. 54). Governance requires widening resources and generating collective learning contexts to develop and share a collective strategic vision of territorial development. Naturally, the creation of territorial contexts of governance in border territories assumes particularities, the development of cooperative relationships, and a dialogue between companies, institutions and the political-administrative authorities of different countries.

Cross-border regions represent a special case in which the overall strategy to improve the competitiveness of the region must address issues related to helping the environment and promoting objectives to firms directly. The dense institutional and policy networks that support cross-border cooperation have not automatically resulted in the establishment of new public-private alliances to address regional and local development issues (OEDC, 2005, p. 15).

5.2 Methodology

To analyse the system of governance in the Raia Central Ibérica, we used an inquiry that was applied to 169 different companies in five sub-regions (three Portuguese and two Spanish) of the area.

5.2.1 The region: The Raia Central Ibérica

The Raia Central Ibérica (RCI) covers the Portuguese sub-regions (NUTs III) of the Interior Centre Region (a great part of Beira Interior): North Beira Interior (BIN) (Almeida, Celorico da Beira, Figueira de Castelo Rodrigo, Guarda, Manteigas, Meda, Pinhel, Sabugal and Trancoso), South Beira Interior (BIS) (Castelo Branco, Idanha-a-Nova, Penamacor and Vila Velha de Ródão) and Cova da Beira (CB) (Belmonte, Covilhã, and Fundão) (Portuguese Raia Central – RCP), and the totality of the Spanish provincial territories of Salamanca and Cáceres situated in the Autonomous Communities of Castilla Y Léon and Extremadura, respectively (Spanish Raia Central – RCE).

The sub-regions of the RCI have a similar social-economic reality: declining populations from a quantitative and qualitative point of view: flaws in the corporate composition; poor economic capacity; and a peripheral geographical and political situation. In general, the sub-regions of the RCI present a disadvantageous situation with a fragile economy and productive performance, making it difficult to create wealth. Their productivity and GDP per capita are about 70 per cent of that of their own countries. The reduced corporate efficiency, associated with low levels of innovation and an unqualified labour force, is partly compensated by the growing employment rate which is superior to the growing population rate.

5.2.2 Gathering information and sample

The main source of data results from an inquiry addressed to different companies selected from the five sub-regions of the RCI. Companies from all sectors, whatever the legal form, with headquarters in the Portuguese Raia Central (RCP) and the Spanish Raia Central (RCE) regions and with more than 10 workers were selected. The sample distribution of the different actors of the Raia Central Ibérica[5] is presented in Table 5.1.

5 Note that, before applying these conditions the number and distribution of the companies was as follows: 699 companies of the RCP and 696 companies of the RCI (237

Table 5.1 Summary of companies studied in the RCI

	RCP			RCE	
	BIN	BIS	CB	Salamanca	Cáceres
Companies (N°)	38	31	36	34	30
Percentage of Total of Companies (%)	14	16	13	13	7
Companies (N°)	105			64	
Percentage of Total of Companies (%)	15			9	
Percentage of Agro Industries (%)	15			8	
Percentage of Commercial Agro-Companies	7			8	

The study includes 169 companies, 33 of which are related to the food chain (13 per cent are agro-industries and 7 per cent are commercial companies producing agro- and food products).

5.2.3 Hypotheses

In this chapter, we analyse the satisfaction of companies of the RCI in terms of governance systems and verify which factors or environmental conditions are associated with higher satisfaction. For this propose, we consider the following set of variables: Local Culture; Local Value Systems; Relationships; Culture and Leisure; Geography and Landscape; Local Labour Offer; Local Labour Offer of Qualified Workers; Academic Institutions; Social Approval; Professional Skills; Openness to Information Circulation; Local Consumers' Knowledge; Local Business Structure; Local Dynamics; Security; Health; Environment and Traffic; Local Road Network; Telecommunication Infrastructures and Distribution Channels Available.

In line with the literature review, we present three hypotheses. The first aims to consider the influence of some contextual variables on the levels of satisfaction of the entrepreneurs in sub-regions RCP and RCE with governance systems. Hypothesis 1 is as follows:

- H1. The levels of satisfaction with governance differ between the RCP and the RCE.

The second hypothesis of the paper analyses the influence of entrepreneurs' perception of problems on the basis of the results of levels of satisfaction with the governance systems in the sub-regions RCP and RCE.

---

companies of the BIN, 193 companies of the BIS, 269 companies of the CB, 268 companies of the Salamanca Province and 428 companies of the Cáceres Province).

According to Domingues (1998, p. 44), territorial governance is 'all the system of relations between institutions, organizations and individuals, which assures the collective choices and their accomplishment'. Thus, the level of sophistication of the relationships between the different actors (public-private, private-private, or public-public partnerships) will vary according to the levels of utilization of ICTs by the companies. Considering both these elements, a second hypothesis is established as:

- H2. The two regions present different levels of problems:
  - H2a. Perceptions of problems differ.
  - H2b. Levels of ICT and technology use differ.

The financial support derived from the local and central administration or from the European Union may influence the different levels of entrepreneurial satisfaction with governance systems. According to Costa (2002, 2003), Bramanti (1999), Deryche (2002), Guillaume (1998), Braczyk et al. (1998), and Cooke (2003), the mechanisms of governance may limit the territorial process of innovation. The territorial contexts of governance assume particularities, mainly by encouraging the development of cooperation relationships and the dialogue between different actors. This leads to the following hypothesis:

- H3. Satisfaction with governance systems varies with type of local initiative.
  - H3a. The level of satisfaction varies with the type of state support.
  - H3b. The level of satisfaction varies with the type of innovative practices.
  - H3c. The level of satisfaction varies with the type of actors involved.

5.3 Data treatment and results

In order to study the governance system in the RCI, we analyse the entrepreneurs' opinions regarding their region, by means of a set of variables[6] (see Table 5.2) that embrace several dimensions and relevant territorial conditions that favour the performance of different sectors.

Applying factor analysis to this set of variables, we obtain the following factors: Localization, Resources, Professional Skills, Dynamism, Environment, and Infrastructures. Table 5.3 presents how satisfied entrepreneurs are with these factors and identifies the components of the governance system, which explain 60 per cent of the variance with a KMO[7] of 0.75. Subsequently, we analyse

6 These variables were classified in a 0 to 3 scale: 0= Not Satisfactory; 1= Low Satisfaction; 2= Medium Satisfaction, and 3= High Satisfaction.

7 The KMO is a statistic that measures the value of adequacy and evaluates the correlations; according to Hill and Hill (2002), a figure of KMO inferior to 0.5 is

Table 5.2 Variables of satisfaction of entrepreneurs

Rotated Components Matrix

	Localization	**Resources**	**Personal**	**Dynamics**	**Environment**	**Infrastructures**
Local Culture	0.84	0.14				
Local Value Systems	0.81	0.17				
Relationships	0.64		0.20		0.11	0.13
Culture and Leisure	0.56	0.34		0.21		-0.22
Geography and Landscape	0.51		0.26	-0.38	0.14	-0.12
Local Labour Offer	0.18	0.78	0.13	-0.15		
Local Labour Offer of Qualified Workers	0.12	0.73	0.15	-0.11		
Academic Institutions	0.12	0.70		0.26	0.13	
Social Approval	0.22		0.84	0.20	0.14	
Professional Skills			0.80	0.16	0.25	0.15
Openness to Information Circulation	0.19	0.40	0.50			0.16
Local Consumers Knowledge				0.67		-0.12
Local Business Structure			0.27	0.58		0.22
Local Dynamics	0.26	0.36	0.28	0.45		0.15
Security			0.36	-0.12	0.78	
Health	0.14	0.45		0.27	0.64	
Environment and Traffic	0.36	-0.13		-0.26	0.46	0.20
Local Road Network		-0.11		0.12	0.35	0.71
Telecommunication Infrastructures		0.17	0.19			0.67
Distribution Channels Available		0.39	0.33		-0.16	0.49

the differences in satisfaction concerning the governance systems of Portugal and Spain (particularly the RCP and the RCE) using the Chi-Square test and verifying the influence of the contextual variables connected to the entrepreneurs' satisfaction, concerning the local governance system, the level of ICT use, and the managements' skills in terms of problem perception on both sides of the border.

Applying factor analysis to this set of variables, we obtain the following factors: Localization, Resources, Professional Skills, Dynamism, Environment, and Infrastructures. Table 5.3 presents how satisfied entrepreneurs are with these factors and identifies the components of the governance system, which explain 60 per cent of the variance with a KMO of 0.75. Subsequently, we analyse the differences in satisfaction concerning the governance systems of Portugal and Spain (particularly the RCP and the RCE) using the Chi-Square test and

unacceptable for factor analysis, a figure equal or superior to 0.9 is excellent, and a figure of 0.8 is good.

Table 5.3 Factor analysis: satisfaction of entrepreneurs

Total Variance Explained			
	Rotation Sums of Squared Loadings		
Component	**Total**	**% of Variance**	**Cumulative %**
1- Localization	2.72	13.58	13.58
2- Resources	2.54	12.70	26.28
3- Professional Skills	2.18	10.92	37.20
4- Dynamism	1.54	7.70	44.90
5- Environment	1.53	7.67	52.57
6- Infrastructures	1.47	7.33	59.90
Kaiser-Meyer-Olkin Measure			0.747638581
Bartlett's Test of Sphericity		Approx. Chi-Square	941.464338
		Df	190
		Sig.	0.000

Table 5.4 Comparison of the levels of satisfaction between the RCP and the RCE

Chi-Square Results	**X^2**	**Prob.**	**Observations**
External Road Network	12.05	0.07	V
Local Road Network	12.20	0.07	V
Infrastructure of Telecommunications	3.11	0.38	X
Distribution Channels Available	6.18	0.10	X
Business Structure	0.79	0.85	X
Local Consumers Knowledge	1.18	0.76	X
Academic Institutions	16.32	0.01	V
Local Labour Offer	6.73	0.08	V
Local Labour Qualified Offer	8.17	0.43	X
Openness to Information Circulation	12.57	0.06	V
Professional skills	6.92	0.08	V
Social Approval	1.24	0.74	X
Security	7.86	0.05	V
Health	37.15	0.00	V
Environment and Traffic	4.28	0.23	X
Culture and Leisure	31.87	0.00	V
Geography and Landscape	1.48	0.48	X
Local Value Systems	4.62	0.20	X
Local Culture	3.55	0.32	X
Relationships	3.60	0.31	X
Local Dynamics	8.30	0.04	V

verifying the influence of the contextual variables connected to the entrepreneurs' satisfaction concerning the local governance system, the level of ICT use, and the managements' skills in terms of problem perception on both sides of the border.

Table 5.4 shows the results of this test, considering differences in the satisfaction levels between the two sub-regions regarding the local governance system. The results allow us to conclude that the aspects concerning the telecommunication infrastructures, distribution and commercialization channels, corporate composition, consumers' claims, manpower supply with the required qualifications, social accomplishment, environment and traffic congestion, geography and landscape, local value system, local culture and sociability, do not demonstrate any differences between the RCP and the RCE regarding the levels of the entrepreneurs' satisfaction.

On the other hand, entrepreneurial satisfaction in terms of external road system, internal road system, superior education institutions, manpower supply, openness/transparency in the circulation of information, professional fulfilment, security, health, culture, leisure, and dynamism differed between the RCP and the RCE. However, note that the Spanish entrepreneurs are generally more satisfied than the Portuguese with the different dimensions of local governance.

An analysis of the general characteristics of the company in terms of the manager's qualifications and the ICT use allows us to conclude that the manager's qualifications, the fact that the companies are connected to the Internet, accomplishing operations of electronic commerce, and use of ICTs to relate with clients/suppliers are not factors that demonstrate differences between the RCP and the RCE (see Table 5.5). Nonetheless, whether the company has a Web Page and computable data does show different behaviours by the entrepreneurs in terms of satisfaction on both sides of the border.

Table 5.5 Comparison of the levels of utilization of ICTs between the RCP and the RCE

Chi-Square Results	X^2	Prob.	Observations
Leadership Skills	1.56	0.46	X
Internet Connection	1.24	0.27	X
Web Page	3.12	0.08	V
Informatization of Business Data	3.89	0.05	V
Electronic Commerce Operations	2.04	0.15	X
Use of ICTs in Relations with Clients and Suppliers	0.60	0.44	X

Regarding the perception of problems by the entrepreneurs in the RCP and RCE, the Chi-Square test also demonstrates functioning differences in the Portuguese and Spanish managerial structures (Table 5.6), particularly in terms of lack of economic ability; lack of support by the public entities, resistence to change,

Table 5.6 Comparison of the perception of the problems by the entrepreneurs between the RCP and the RCE

Chi-Square Results	X^2	Prob.	Observations
Lack of Economic Capacity	12.03	0.01	V
Lack of Government Support	11.95	0.01	V
Resistance to Change	18.39	0.00	V
Lack of Cooperation Between Local Agents	19.84	0.00	V
Lack of Qualified Personnel	20.91	0.00	V
Lack of Information on Technology	9.24	0.26	X
Lack of Market Information	8.34	0.04	V
Lack of Education and Training	3.95	0.27	X
Aspects of Local Culture	1.03	0.79	X
Aspects of Local Tradition	1.70	0.64	X

lack of cooperation between local agents, lack of qualified personnel, and lack of information about the markets. However, contextual variables such as lack of information about technology, lack of qualifications/education, aspects of local culture, and aspects related to the local traditions do not show different influences on the perception of problems by the entrepreneurs who operate in the Portuguese or Spanish regions of the RCI.

In the data analysis, we apply cluster analysis methodology (K-means cluster) to verify how the companies evaluate levels of satisfaction with the local governance system. This methodological application was shown to be appropriate and revealed statistical significance for most variables. According to Pestano and Gageiro (2000), the levels of significance of the tests F ($\rho<0.05$) show that each factor has a differentiated contribution in the three groups.

The cluster analysis identifies three groups of companies regarding entrepreneurs' satisfaction with the region's characteristics (Table 5.7). The first cluster, consisting of 86 companies, shows reduced results in terms of satisfaction. The second cluster, consisting of 40 companies, presents medium satisfaction. The third cluster, consisting of 42 companies, demonstrates the best results in terms of satisfaction with the region's characteristics and the governance system.

As for financial support from the local administration, central administration, or the European Union, only the funds from the local administration influence the different levels of satisfaction. In effect, cluster 3 stands out from clusters 1 and 2 (see Table 5.8).

Regarding the previously rated clusters, we have verified that the entrepreneurs from the more satisfied cluster also have a major percentage of companies that introduce innovations mainly, at the product level. As for the entrepreneurs' susceptibility concerning the existence of a collective learning effect in the region and the dissemination of know-how (exchange of knowledge, cooperation among agents, innovation dissemination), the previously defined groups of companies

Table 5.7 Cluster analysis

Final Cluster Centres	1 (n=86)	2 (n=40)	3 (n=42)
External Road Network	1.0	1.6	1.6
Local Road Network	1.0	1.5	1.4
Infrastructure of Telecommunications	1.3	1.8	2.0
Distribution Channels Available	1.1	1.5	1.8
Business Structure	0.8	1.1	1.5
Local Consumers Knowledge	2.0	1.7	2.2
Academic Institutions	1.0	1.6	2.5
Local Labour Offer	0.6	1.3	2.0
Local Labour Qualified Offer	0.5	1.0	1.6
Openness to Information Circulation	0.6	1.2	1.7
Professional skills	0.7	1.4	1.9
Social Approval	0.9	1.4	2.0
Security	1.5	2.1	2.3
Health	0.6	1.6	2.2
Environment and Traffic	1.9	2.0	2.3
Culture and Leisure	1.3	1.4	2.1
Geography and Landscape	2.4	2.6	2.7
Local Value Systems	1.5	2.0	2.4
Local Culture	1.7	1.9	2.5
Relationships	1.3	1.7	2.2
Local Dynamics	0.9	1.3	2.0

Table 5.8 ANOVA – Group differences considering state support received

		Sum of Squares	Df	Mean Square	F	Sig.	
Local Administration	Between Groups	0.42	2	0.21	2.61	0.08	1,2<3
	Within Groups	13.24	165	0.08			
	Total	13.66	167				
Central Administration	Between Groups	0.11	2	0.05	0.33	0.72	
	Within Groups	26.41	165	0.16			
	Total	26.52	167				
EU Funds	Between Groups	0.44	2	0.22	1.03	0.36	
	Within Groups	35.08	165	0.21			
	Total	35.52	167				
EU Initiatives	Between Groups	0.01	2	0.01	0.49	0.61	
	Within Groups	1.96	165	0.01			
	Total	1.98	167				

also present some differences: about 33 per cent of the entrepreneurs from cluster 3 say that there is a learning effect in the region as opposed to the other clusters of companies with minor local satisfaction. These results also reflect the fact that the entrepreneurs do not feel that there is a common identity in the studied region. In terms of local identity, even the most satisfied cluster includes differences from the other clusters: only 14 per cent of the entrepreneurs say that there is a common local identity (Table 5.9).

Table 5.9 Innovation and learning

Percentages	Cluster Number of Case		
	1	2	3
Company Innovations	60.5	47.5	73.8
Product Innovation	47.7	37.5	59.5
Process Innovation	25.6	20.0	23.8
Organizational Innovation	25.6	20.0	23.8
Collective Local Learning Effect	22.1	7.5	33.3
Non Collective Local Learning Effect	30.2	45.0	14.3
Company Cooperative Arrangements	54.7	47.5	61.9

The governance system involves a set of institutional actors who have the ability to make decisions in the territory. It has a central role in the territorial process of innovation through the projects that it defines, the regional politics, regulation of local activities, and engagement modalities. The governance of a territory must function through local actors (with interaction, shared culture and history, and development of trust in the local communities), cooperation networks (formal or informal), and in a trans-territorial and systemic perspective.

Regarding this, we realize that cooperation networks are not privileged. However, the group of entrepreneurs who are most satisfied with the governance system includes a major percentage of companies that establish cooperation with local actors.

Concerning the importance of cooperation to innovate conferred by the different local actors (Table 5.10), there are significant differences among clusters regarding the local actors, concerning the level of cooperation with companies from the group, the level of suppliers, competitors, and local public administration (highlighting the most satisfied group from the other clusters, except in terms of competitors). Nevertheless, the entrepreneurs place the same value on cooperation with clients, consultants, R&D institutions, and laboratories, institutions of higher education, development associations, managerial associations; central public administration.

Table 5.10 ANOVA – Importance of different actors to the innovation process

	Sum of Squares	df	Mean Square	F	Sig.	
Companies of the Group	14.85	2	7.42	4.17	0.02	1,2<3
Suppliers	10.00	2	5.00	2.86	0.07	1<2,3
Clients	3.76	2	1.88	1.13	0.33	
Competitors	4.87	2	2.43	4.01	0.02	1,3<2
Consultants	2.10	2	1.05	0.85	0.43	
R&D Institutions	1.87	2	0.93	0.58	0.56	
Academic Institutions	2.65	2	1.32	0.78	0.47	
Development Associations	0.75	2	0.37	0.42	0.66	
Business Associations	5.26	2	2.63	1.58	0.22	
Central Administration	4.32	2	2.16	1.99	0.15	1,2<3
Local Administration	8.32	2	4.16	4.17	0.02	1,2<3
Workers Associations	0.67	2	0.33	1.56	0.22	

5.4 Final reflections

The economic literature on territorial governance suggests that one of the essential elements in the analysis of the system of territorial governance is the present actors' evaluation of the performance of the entities responsible for its governance. The assembled data and the application of the described methodology indicates that the evaluation of the system of territorial governance is related to the satisfaction of the entrepreneurs of the Raia Central Ibérica, and that there are differences in the satisfaction level between the sub-regions of the RCP and the RCE; the Spanish entrepreneurs generally show more satisfaction than the Portuguese: namely, in terms of the road system, institutions of higher education, manpower supply, openness in the circulation of information, professional fulfilment, security, health, culture, leisure, and the region's dynamism.

The lack of economic ability, lack of support from public entities, resistance to change, the lack of cooperation among local agents, lack of qualified personnel, lack of information about technology, and lack of information on the markets also influence satisfaction regarding the local governance system from both sides of the border.

The financial support from the local administration is shown to influence the level of entrepreneurs' satisfaction. Their satisfaction with the governance system mainly occurs at the level of introducing innovation with regard to the existence of a collective learning effect in the region and the dissemination of know-how and of a common identity in the region and in terms of cooperation. Nonetheless, the relationships with the cooperation networks are not privileged in the region through a collective model of functioning and economic development.

There were some limitations in this study, which indicates the need for further analysis, particularly a wider sample of, and extended analysis to include, other actors present in the territory, such as institutions that provide support and assistance to the managerial activity; technological centres; managerial and development institutions; education, training, and R&D systems; universities and polytechnics; and training institutes and technological schools. There is also a need to widen the research to other regions.

References

Arndt, C. and Oman, C. (2006), *Uses and Abuses of Governance Indicators*, Paris, OECD Publications.

Benko, G. and Lipietz, A. (eds) (1992), *Les Régions qui Gagnent*, Paris, PUF.

Bogason, P. (2000), *Public Policy and Local Governance: Institutions in Postmodern Society*, Cheltenham, UK, Edward Elgar Publishers.

Braczyk, H-J., Cooke, P. and Heidenreich, M. (eds) (1998), *Regional Innovation System,* London, UCL Press.

Bramanti, A. (1999), 'From Space to Territory: Relational Development and Territorial Competitiveness', *Revue d'Economie Régionale et Urbaine*, vol. 3, pp. 633–654.

Brenner, N. (2004), *New State Spaces: Urban Governance and the Rescaling of Statehood,* Oxford, Oxford University Press.

Cabus, P. (2003), 'Governance in Flanders' Regional Policy: Subregional Platforms as Development Coalitions', Paper presented at *Regional Science Association International Conference,* 12–15 April, Pisa.

Castels, M. (2000), *The Rise of the Network Society*, Oxford, Blackwell Publishers.

CEC (2002) *Report from the Commission on European Governance*, Commission of the European Communities (CEC), COM (2002) 705 final.

Cooke, P. (2003), Strategies for Regional Innovation Systems: Learning Transfer and Applications. *Policy Papers*, Vienna, United Nations Industrial Development Organization (UNIDO).

Costa, P. (2002), *Efeitos de Meio num Bairro Cultural: Alguns Resultados de um Estudo na Zona Bairro Alto- Chiado*, Paper presented at the IX Encontro da APDR: A nova Economia e Desenvolvimento Regional, 27–29 de Julho, Lisboa.

Costa, P. (2003), *A Fileira da Cultura e a Competitividade Territorial: o caso da Área Metropolitana de Lisboa*, Unpublished Doctoral Dissertation, ISEG, Portugal.

Crouch, C., Le Galés, P., Trigilia, C. and Voelzkow, H. (2001), *Local Production Systems in Europe: Rise or Demise,* Oxford, Oxford University Press.

Deryche, P-D. (2002), *Analyse de Système et Prospective Régionale – Histoire et Développements Récents*, Paper presented at the 38iéme Colloque ASRDLF, 21–23 August, Trois Rivières, Canada.

Domingues, A. (1998), *Universidade, Empresas e Território: Reflexões em Torno de uma Articulação (Im)possível*, Paper presented at the APDR on the Ensino, Empresas e Território, 18–20 June, Coimbra.

European Commission (2001), *European Governance, White Paper*, European Commission, Brussels.

Fermisson, J. (2005), *Das Estratégias dos Actores à Estratégia do Território – O Papel dos Actores Locais de Governância Face ao Processo de Mundialização*, Unpublished Master's Dissertation, Faculdade de Ciências Sociais e Humanas da Universidade Nova de Lisboa, Lisboa, Portugal.

Gualini, E. (2004), 'Integration, Diversity, Plurality: Territorial Governance and the Reconstruction of Legitimacy in a European "Postnational" State', *Geopolitics*, vol.9, nº 3, Autumn, pp. 542–563.

Guerra, I. (2000), 'O Planeamento Estratégico das Cidades: Organização do Espaço e Acção Colectiva', *Centro de Estudos Sociais Cidades: Comunidades e Territórios*, 1, Lisboa: Universidade de Lisboa, CES.

Guillaume, H. (1998), *Rapport de Mission sur la Technologie et L'Innovation.* Ministère de L'Economie et de L'Industrie from http://www.finances.gouv.fr/innovation/guillaume/hg27.htm

Healey, P. (1997*), Collaborative Planning. Shaping Spaces in Fragmented Societies*, London, Macmillan.

Hendricks, F. (2006), 'Shifts in Gouvernance in a Policentric Urban Regions: The case of the Dutch Randstad', *International Journal of Public Administration* (forthcoming).

Henriques, J.M. (1985), Planeamento Regional como Diálogo: uma Proposta de Conceptualização. *Revista de Estudos Urbanos e Regionais*, 3, Edições Afrontamento, Porto, pp. 121–127.

Hill, M.M.and Hill, A, (2002), *Investigação por Questionário*, Edições Silabo.

Le Galés, P. (2003), *Le Retour des Villes Européennes*, Paris, Presses de Sciences Politiques.

Lopes, R. (2001), *Competitividade, Inovação e Territórios*, Oeiras, Celta Editora.

Michalski, W. and Miller, R. and Stevens, B. (2002). Governance in the 21st Century: Power in the Global Knowledge Economy and Society, in OECD (ed.), *Governance in the 21st Century*, Paris, OECD.

Natário, M. (2004), *Inovação, Competitividade e Demografia Empresarial: O caso da Raia Central Ibérica.* Unpublished Doctoral Dissertation, Universidade de Évora, Portugal.

Neto, P. (1999), 'O Portfolio Relacional dos Territórios na Reformulação das Vantagens Comparativas Inter-Territoriais', in: APDR (ed.), *Emprego e Desenvolvimento Regional.* Actas do V Encontro Nacional da APDR, Vol. 2, Coimbra, Colecção APDR.

OECD (1993), *Partnerships: The Key to Job Creation, Experiences from OECD Countries*, Paris , OECD Publications.

OECD (1998), *Local Management for More Effective Employment Policies*, Paris, OECD Publications.

OECD (1999a), *Best Practices in Local Development*, LEED Notebook 27, Paris, OECD Publications.

OECD (1999b), *Managing Accountability in Intergovernmental Partnerships*, Unclassified Document [PUMA/RD(99)4/ FINAL], Public Management Services, Paris, OECD Publications.

OECD (1999c), *Study on Local Partnerships. Project Methodology*, Official Document [DT/LEED/DC(99)1/REV1], LEED Programme, Territorial Development Service, Paris, OECD Publications.

OECD (2000a), *Urban Renaissance: Belfast's Lessons for Policy and Partnerships*, Paris, OECD Publications.

OECD (2000b), *The Creative Society of the 21st Century*, Paris, OECD Publications.

OECD (2001a), *Local Partnerships for Better Governance*, Paris, OECD Publications.

OECD (2001b), *Governance in the 21st Century*, Paris, OECD Publications.

OECD (2001c), *Cities for Citizens: The Role of Metropolitan Governance*, Paris, OECD Publications.

OECD (2004), *New forms of Governance for Economic Development*, Paris, OECD Publications.

OECD (2005), *Building Competitive Regions – Strategies and Governance*, Paris, OECD Publications.

OECD (2006a) *Uses and Abuses of Governance Indicators,* Paris, OECD Publications.

OECD (2006b) The New Rural Paradigm – Policies and Governance, Paris, OECD Publications.

Pestana, M.J. and Gageiro, J.N. (2000), *Análise de Dados para as Ciências Sociais. A Complementaridade do SPSS*, 2ª Edição, Edições Sílabo.

Peters, B.G. (2001), *The Future of Governing*, Lawrence, University Press of Kansas.

Pollit, C. and Bouckaert, G. (2004), *Public Management Reform : a Comparative Analysis*, Oxford, Oxford University Press.

Salet, W. (2006), 'Rescaling Territorial Governance in the Randstad Holland: The Responsiveness of Spatial and Institutional Strategies to Changing Socio-economic Interactions', *European Planning Studies*, vol. 14, n° 7, August, pp.959–977.

Storper, M. and Harrison, B. (1991), 'Flexibility, Hierarchy and Regional Development: the Changing Structure of Industrial Production Systems and their Forms of Governance in the 1990's', *Research Policy*, vol. 20, pp. 407–422.

Storper, M. and Harrison, B. (1992), 'Flexibilité, hiérarchie et dévelopment régional: les changements de structure des systèmes productifs industriels et leurs nouveaux modes de gouvernance dans les années 1990', in: G. Benko and A. Lipietz (eds.), *Les Régions qui Gagnent.* Paris: PUF.

Tatenhove, J. van and Leroy, P. (2000), 'The Institutionalization of Environmental Policies', in: J. van Tatenhove, B. Arts and P. Leroy (eds), *Political Modernization and the Environment. The Renewal of Environmental Policy Arrangements*, Dordrecht, Kluwer Academic Publishers, pp. 17–33.

Voets, J. and Rynck, F. de (2006), 'Rescaling Territorial Governance: a Flemish Perpective', *European Planning Studies,* vol. 4, n° 7, August, pp. 905–922.

Chapter 6
The Role of Cultural Values in Rural Development

Edina Szlanyinka

Introduction

Tourism has become a development tool for many rural and more isolated areas to supplement traditional industries that are often in decline because tourism is an economic activity involving the exchange of billions of dollars each month – a social science to be analysed, trends to be identified, and costs/benefits to be computed (Lundberg et al., 1995).

However tourism is not just an economic activity but rather a complex social and cultural phenomenon and tourism trends more and more reflect this complexity as it was defined at the Conference of Culture and Patrimony Interreg as follows: 'The demands of tourism are now more diversified, less attached to seasonal variation and more challenging; and they are also affected to a greater extent by cultural and artistic factors' (Conferencia de Ciudades del Arco Atlantico – Cultura y Patrimonio Interreg III 2002).

Cultural heritage represents the record of mankind's achievements and relationships with the world. Therefore, it always has a local dimension, though sometimes it embeds universally shared values. Labelling something as heritage represents a value judgment, which distinguishes that particular object from others, adding new meaning to it. Cultural heritage summarizes people's identities, shapes those of communities, and to this extent contributes to the creation of social capital.

Heritage is a social, economic, and cultural resource. Heritage valuation becomes a tool to better understand the significance of heritage to different sections of society (Riganti and Nijkamp, 2004).

Navrud and Ready (2002) realized that economic development threatens many of the world's cultural and historical treasures (the assignment of economic values to cultural and historical resources). The assignment of values is not straightforward since many cultural and historical resources are difficult to allocate by the market mechanism. The assignment of values is critical. When compared with the economic value of economic development, cultural and historic resource market values often appear lacking.

Market economies have generally proved to be capable of allocating goods, services, and other resources in an efficient manner. One exception is public goods

– those goods that are non-rival and non-excludable in consumption, providing little incentive for consumers and firms to develop a market. Examples of these goods include national defence, environmental quality, and cultural and historical resources (Whitehead, 2002).

The valuation process aims to assess existing values as attached by the relevant population. On the other hand, the ultimate aim in the context of policy analysis is to value in order to achieve the valorization of our heritage, in other words, to add new values to the existing ones. For that reason, valuation represents a crucial step in the management of cultural heritage and in regional development. The specific nature of cultural heritage as a collective good also implies that the investment and maintenance costs have to be covered by all citizens (Riganti and Nijkamp, 2004).

Rural communities and peripheral areas face the challenge of continuous economic development. Where primary traditional industries are in decline, tourism often becomes another tool to help create jobs and to raise the standards of living (Fleischer and Felsenstein, 2000; Hill, 1993; Sharpley et al., 1997). These areas realize this potential through the development of local resources, culture, and heritage (Kochel, 1994; Lewis, 1998). The integration of such alternative sources may help to sustain local economies and to encourage local development (Prohaska, 1995). Actually, many tourists seek rural destinations which offer pleasant experiences related to the natural environment, historic heritage, and cultural patterns (Butler and Hall, 1998; WTO, 1994). It is this culture and heritage that are often well preserved between generations in rural areas, and it is in periods of economic decline that their residents seem to cling more to a distinct heritage. Culture and tourism then become resources for socio-economic development in rural and peripheral communities. (Jolliffe and Baum, 1999).

The concepts of culture, rural, and tourism are multidimensional and interrelated. Fredericks (1993) discusses tourism as building on perceived and existing local amenities like historical sites, natural beauty, and clean air, all similar to those in rural areas. Hardy (1988) and Millar (1989) refer to heritage tourism as cultural traditions, places, and values that groups conserve. Similarly, McNulty (1991) and Weiler and Hall (1992) consider culture to include family patterns, folklore, social customs, museums, monuments, historical structures, and landmarks. Others include wilderness areas, valued landscapes, natural history, buildings, and artefacts as part of cultural tourism (Prentice, 1993; Tassell and Tassell, 1990)

In the same way, we can also add to the cultural traditions those of food, with the example of the Slow Food Organization, which represents important events (Cheese and Salone del Gusto – Italy) during which the producers meet the consumers: the former have the chance to show their wares and to promote them, the latter to taste the products and to gain information about them. Also involving the public authorities of the food territory of origin, the local associations and consortia, and the distributors, the fairs are further opportunities to exchange information and to facilitate the building up of cooperation or business relationships among different operators (Nosi and Zanni, 2004).

Studying gastronomic festivals, food-related events and fairs one of the main tasks to better understand them is to determine the motives of visitors. 'A motive is an internal factor that arouses, directs, and integrates a person's behaviour (Iso-Ahola, 1980). A decision to visit a festival is a directed action which is triggered by a desire to meet a need. Although they are only one of multiple variables that explain behaviour (others would include learning, cultural conditioning, social influences, and perceptions), they are the starting point that launches the decision process.

The tourism literature has long recognized that a pleasure trip is rarely the result of a single motive. Tourists' motives are likely to be multiple (Crompton, 1979; Mansfeld, 1992; Pearce, 1982; Uysal et al., 1993). This multiplicity may occur at both the individual and the aggregate levels of analysis. At the individual level, a visitor may have several different needs which he/she desires to satisfy through a festival visit. For example, a need to interact with the family, often inhibited by the independent actions of individual members in the home environment, may be accompanied by a desire for cultural enrichment. The satisfaction of these needs may be facilitated by different elements in the package of offerings which constitutes a festival. At the aggregate level of analysis, different visitors may engage in the same package element and derive different benefits from the experience. Thus, for some, eating a particular food may be a means of experiencing a cultural tradition, while for others it may be a means of facilitating inter- or intra-group socialization.

Organizers have multiple reasons for staging festivals, and identifying the visitor benefits that are sought provides an answer to the question: 'Are the rationales for staging a festival consistent with the benefits that visitors receive from it?' (Mayfield and Crompton, 1995). Finally, identifying and prioritizing motives is a key ingredient in understanding visitors' decision processes (Crompton and McKay, 1997).

In Hungary the new trend of rural development is the mobilizing the use of local cultural and human resources, as well as the traditional material factors generally considered. Gastronomic culture plays an important role in the conception of a cultural economy and rural development.

Gastronomic tourism is becoming increasingly important; its attractions have a great interest for all levels of society. The food is no longer looked upon as just a necessity for survival, but also as a means of enriching our experiences, expressing our personal identities; in other words, adding to the general quality of life. The process of globalization has given rise to the interest in, and focus on, regional identities and the roots of our culture.

This chapter primarily focuses on the role of culture in our society and secondly draws links between rural development and gastronomic tourism: the opportunities that gastronomy can offer in the development of rural areas, and the possibility of maintaining the valuable heritage brought through local gastronomy. Mainly concentrating on Hungarian cultural values, cultural heritage, putting emphasis on gastronomy and gastronomic festivals, I have been carrying out research in Hungary on the impacts of festival tourism, first and foremost gastronomic festivals, and I have been surveying a collection of gastronomic festivals from all over the world based on Internet research. This list of festivals enables me to explore the motivation behind gastronomic festivals, and also enables me to classify different types of festivals.

The following analysis covers the roles of gastronomic festivals, how they can mobilize the community, strengthen identity, bring people together, preserve our cultural past by bringing it into the modern reality and teach children to respect culture. These events play important roles in the life of people. By collecting and studying the impacts of gastronomic tourism, we are capable of enriching the rural development programmes.

6.1 Traditions of gastronomy

In recent years, food and wine have played an increasingly more important role in tourism. Not only are they featured in tourism promotions, but food and wine tourism has become a significant part of tourism in general in the past few decades.

Celebrations have always been good occasions for eating and drinking for Hungarian people. We drink a toast when a child is born, at wedding ceremonies, pig slaughtering, wine harvests, and even at burial feasts. In the life of craftsmen, the initiation of new masters or servants was always accompanied by a feast. Similarly, a sales transaction was valid only if it was confirmed by a good drink. All this can be traced back to the rise of the monasteries, which came into existence together with the surrounding villages. The croplands, fruit gardens and vineyards were cultivated regularly. The Hungarians learned how to grow different vegetables from the Slavic servants in the monasteries, and the local inhabitants gradually adopted the cultivation of vineyards in the Transdanubian region. A large number of villages, also started to make a living from bee-keeping. In the time of Saint Stephen, the first Hungarian king, large herds of partly domesticated, partly wild pigs were kept all around the country. Along with the economic changes, ways and habits of eating also changed (http://www.hungary.com).

Hungarian dining rests on the pillar of wine, as do the French, the Italian and the Spanish culinary traditions. The first written record on the subject is dated 1484: 'Hungarians have a great deal to eat and drink at their feasts, and they drink many kinds of wine. (The same habit was attributed by the historians to Galienus, Roman Emperor.) ... At the most sumptuous feasts, several different types of wine are served.'

6.1.1 The history of Hungarian wine

'Hungarian wines are the (somewhat) hidden gems among Europe's greats' is a quotation found when searching the Internet, from an article entitled *Not Just Known for Goulash: A Tour of Hungarian Wines* by Ursula Furi-Perry (2005). She introduces the reader to Hungary's over 100,000 hectares of land devoted to viniculture and wine production, as well as 22 distinct wine regions, home to some excellent – 'yet largely undiscovered' – wines, which can be called in Hungarian: *bor*.

These 22 wine production areas are located in four large regions. The traditions of wine production which have developed over a long period are characteristic of them all, and an important part of Hungary's national heritage. The position of the growers, favourable or adverse, and the quality and reliability of Hungarian wines have always reflected the political conditions in the country.

Wine fulfilled a significant role in the life of the ancient Hungarian tribes, both as a healthy drink and as a symbol of faith and ritual. Travellers in those times and certain records, including the Byzantine encyclopaedias, mentioned the great number of vineyards in the original homeland of the Hungarians, for whom the most important drinks were milk and wine.

In those days, as also later in the Middle Ages, agreements of any consequence were sealed and sanctified by the parties with a drink of wine. The main witness was in charge of blessing the wine. The custom and ritual of a pact consecrated with blood was described by Herodotus with respect to peoples related to Hungarians, and by certain Byzantine records in connection with the Hungarians themselves. We can infer that the blood treaty, which fused the tribes into the unified Hungarian nation, was also accompanied by a solemn and momentous blessing with wine.

Of all the living European languages, only Greek and Hungarian have a truly native word for wine. The other languages all use a derivative of the Latin noun 'vinum'. These circumstances suggest that some of the ancient Hungarian grape growing and winemaking traditions go back to the peoples who inhabited the native land of *vitis annifera*, the wine-producing grape.

Once Hungarians had settled the land, they took up Christianity. As a result, wine production became ever more important, and growing areas were expanded in Hungary. The missionaries brought a number of innovations to growing and vinification practices.

On the one hand, dealing with wine relied on the ancient traditions Hungarians had brought with them from the Caucasian Mountains, deep in Asia. On the other hand, they had the local version of the Latin knowledge of wine in their land, especially around Lake Balaton. Finally, experience was imported by Benedictine missionaries, and by certain teaching mendicants and immigrant settlers from Italy, Burgundy and the Rhine. This very colourful heritage – combined with extremely varied soils and microclimates – has been instrumental in imparting a unique quality to certain Hungarian wines in well-defined growing locations.

Making wine was an activity that entailed special rights and responsibilities. Serfs were allowed to freely bequeath, buy and sell their vines.

Local wine community governments, then known as Promontorium, were first recorded in 1271. Litigation documents pertaining to the origin of some wines, and especially the Buda Book of Law from the 14th century, attest to the elaborate regulation of the wine hills and a rigorous practice of wine law. An ancient tradition prescribed that 'all proprietors of vines abide by the same law, be they of the nobility, of arms, or of the order of serfs' (www.capvex.com/heritage.shtml).

Over the centuries the territory of Hungary has experienced the Turkish occupation, wars, phylloxera, and irresponsible political regimes which caused serious difficulties for vineyard holders. Today the wine communities have been revitalized, family estates are springing up, and a few cooperatives do exemplary work, offering a good base for making theme routes: namely, wine-roads or wine trails.

6.1.2 Traditional Hungarian dishes

The traditional Hungarian dishes abound in piquant flavours and aromas. Dishes are flavourful, spicy and often rather heavy. Flavours of Hungarian dishes are based on centuries-old traditions in spicing and preparation methods. The exquisite ingredients are produced by local agriculture and husbandry. Paprika, onion and garlic are to be found everywhere.

In the Middle Ages fish soup was the most popular and most loved fish meal in Hungary, and the fish meal could not be absent from King Matthias's table. A cookery book from 1860 contains 400 fish recipes.

The most well-known specialities of Hungarian cuisine such as goulash soup, or 'gulyásleves', the different varieties of stew or 'pörkölt' and 'paprikás' are red with paprika. It should be noted that dishes flavoured with paprika need not be hot. In fact, some of the finest brands of paprika are not strong at all. Fresh green peppers and tomatoes are also frequently used to flavour these dishes, the essence of which is the same: braise any kind of meat in lard/oil with onions, and season it richly with red pepper. Sour cream is an essential ingredient in many Hungarian dishes, too. It is a basic ingredient of 'paprikás' (any kind of meat or just plain potatoes prepared in paprika sauce, to which sour cream has been added), as well as in many Hungarian vegetable dishes and casserole stews.

The use of pasta for desserts like noodles with cottage-cheese, or 'túróscsusza', egg squares with fried cabbage, or 'káposztáskocka' – especially as sweets with sugared nuts, poppy seeds, fruit jellies or jam – is probably peculiar to Hungary.

A real Hungarian speciality, strudel or 'rétes' is a firm favourite even with people who do not like pastry or sweets as a rule. It is no wonder, because its dominant flavour is given by the various fillings, savoury or sweet, while the pastry itself – light and crispy – only enhances their delicacy. For the filling, fresh fruits, such as apples and cherries are used, as well as cottage cheese, cabbage, pumpkin, nuts, rice and poppy seeds (www.cheflaszlo.com/traditionalhungarian.html).

6.2 Characteristics of Hungarian gastronomic tourism

Gastronomic tourism refers to tourism or travel which is motivated, at least in part, by an interest in food and drink. It can be defined as 'travel in order to search for, and enjoy, prepared food and drink' (Wolf, 2002) and includes 'all unique and memorable gastronomic experiences' (Santich, 2004).

In general, gastronomic tourism (food and wine tourism, culinary tourism, cuisine tourism) is a form of tourism focused on food and drink, gastronomic specialities in particular. This can take the form of a live-in cooking school, experiencing traditional gastronomic feasts and celebrations, festivals, helping with the grape harvest, and visiting regional wineries and food producers along a Wine and Food, Gastronomic Trail or Route.

The most popular brochures of the Hungarian National Tourist Office are 'Active Tourism' and 'Wine and Gastronomy' at different exhibitions in European cities. The popularity of wine trails and festivals, gastronomic routes and festivals is growing every year. This shows that people like to have fun, but culinary adventure has an important role, too. There is an opportunity for the rural areas to develop their supply and exploit natural resources.

2006 was the 'year of gastronomy and wine' in Hungary.

There now follows a short description of the four main attractions of gastronomic tourism which are built upon Hungary's cultural heritage:

- Wine roads
- Wine harvests
- Gastronomic roads
- Gastronomic festivals

6.2.1 Hungary's wine roads

The wine roads are an invitation to visitors to follow a predetermined, pre-selected route, which will take them through the most scenic parts of a vine region, including visits to selected wine-making units, give them the opportunity to taste wines typical of each region accompanied by dishes from the local gastronomy, and admire places of natural, archaeological or other interest.

The Villány-Siklós Wine-Route is Hungary's first wine-route, established in the autumn of 1994 on the basis of the endowments of the Villány region, with the joint efforts of eight settlements, private vine-growers and wine-makers, entrepreneurs, and NGOs, following the example of German, Austrian and French models. As the result of the gradual growth of interest over the past years, today the Wine-Route winds through 11 settlements.

The message of the Wine-Route is: get into contact with the traditions and the rural life through quality products (local wine) and quality services, to gain unique experiences, to build human relations, and to enjoy active recreation. In the neighbourhood of the Villány Hills, from the Shrine of Máriagyűd to the Cellar Village of Palkonya, there is a wide range of wine-tasting procedures, traditional cellars, wine contests, festivals, folklore events waiting for the guests. Along the way, the historical memorial sites, the beauties of nature, and the folk architecture in the villages and in the vineyards offer interesting sights. Local dishes and fine wines await the guests in the houses offering rural accommodation, the pensions and the hotels. The quality of the tourism facilities on the Wine-Route is guaranteed by the system of authorized services bearing a trade-mark and the notice boards of the information system erected along the roads. The guests are also helped by the Service Catalogue of the Villány-Siklós Wine-Route using the same logical and graphics system as applied on the direction boards, as well as by the information offices (www.borut.hu/_eng/koszonto.htm).

In the region, various attractions connected to the traditional way of life also play an important role.

6.2.2 Wine festivals, harvests

In Hungary it is not just the wine-routes that are popular, but the wine festivals and harvests too. In the last 10 years, their number has multiplied, although these events are based on Hungarian tradition. The festivals are mainly held in autumn at the end of the grape picking period, but a few take place at other times of the year.

These festivals provide a chance to introduce national producers, national wines, and traditions to Hungarians, foreigners, wine-lovers, tourists and representatives of the profession. They also provide the opportunity to hold conferences, professional meetings, wine competitions and exhibitions. They help to build up the image of the region.

The events, festivals, have to be harmonized with the regional development projects and the individual settlements also have to fine-tune their festivals with each other to avoid the unfortunate situation that once happened in the area of Mount Badacsony situated on the Northern coast of Lake Balaton and famous for its wine, where three important wine-producing townships held their wine festivals at the same time.

It is necessary to have a good marketing and action plan because, if a festival is organized well, producers can meet with consumers who will give them immediate feedback. The festivals provide a good opportunity to advertise. Also, the settlements can introduce themselves, and generally, moderate alcohol-drinking can create a good mood, and help recall the atmosphere of traditional gatherings.

6.2.3 Gastronomic routes

In the last few years, gastronomic roads have appeared in Hungary, and one example will now be described.

In 2001, the Szatmár-Beregi Plum Road Association organized such a route with the purpose of presenting the tourist attractions of the Szatmár-Bereg region focusing on plums, and to introduce the famous plum-pálinka (spirit) and the delicious dishes made of plums in the form of stewed prunes to the visitors. The Association has 18 members such as town councils, plum farmers, rural feast organizers and hostel managers.

For the tourists there are two roads: one of them is 4–5 days long and is called the Big Plum Road, starting and ending in Vásárosnamény (passing through Vásárosnamény, Tákos, Csaroda, Beregdaróc, Beregsurány, Márokpapi, Tarpa,

Table 6.1 Survey of gastronomic festivals, 2003–2006

The websites	The keywords: gasztronómiai fesztivál (Hungarian)		gastronomic festival (English)		food and drink festival (English)		gastronomski festival (Croatian)		гастрономический фестиваль (Russian)	
	Number of search results in the mentioned year									
	2003	2006	2003	2006	2003	2006	2003	2006	2003	2006
www.origo.hu	124	1823	-	13	-	30	-	-	-	-
www.yahoo.com	121	50500	4610	165000	208000	4840000	8	195	2	7570
www.google.com	159	121000	5510	379000	229000	14900000	12	1500	825	21600
www.rambler.ru	-	4	17	1278	536	1838	-	-	727	1284

Source: Own data collection, April 2003 and April 2006.

Tivadar, Penyige, Túristvándi, Szatmárcseke, Tiszacsécse, Milota, Sonkád, Kölcse, Panyola, Vásárosnamény); and the other one takes 1 or 2 days, and is called the Small Plum Road, from Penyige to Panyola (Penyige, Túristvándi, Szatmárcseke, Tiszacsécse, Milota, Sonkád, Kölcse, Panyola).

In 16 settlements, 32 information panels serve as guidance for the visitors.

6.3 Survey of gastronomic festivals

Observing a multitude of web pages since 2003 on the subject related to world gastronomy, it is apparent that many gastronomic events that had been organized in Hungary have now turned into tradition. Table 6.1 represents a survey of gastronomic festivals.

It is clear that since 2003 the number of gastronomic festivals has multiplied. Numerous gastronomic festivals are everywhere, or people mention many times that the gastronomy and the festival lock into each other. What is behind all this, of course, in our globalized world is that we use Internet much more, so every day we get more and more information, but I assume this big change during this 3-year period is not just explained by the Internet globalization. There is in fact a recovery of traditions, recognition of the importance of gastronomy, because, in the words of the old saying: 'we are what we eat'.

6.3.1 Definition, aims and typology of gastronomic festivals

A 'gastronomic festival', as defined here, is an event which is organized year after year for celebrating food (raw products, cooked meals: everything that we can eat) and drink (wine, beer, alcoholic and non-alcoholic beverages: everything that we can drink), where people can meet each other to have a 'culinary adventure': a cooking competition, an exhibition of edible and drinkable products, or just a chance to eat together.

The aims of gastronomic festivals are: bringing people together, not only to create a sense of identity, but also to encourage peaceful social relations too, to have the opportunity to organize professional meetings, exhibitions of products, food, drinks, to exchange information, to learn about culinary methods, cooking tips, to know, to taste traditional products, food, drinks, to feel good, etc. But one important purpose the festival can serve is the possibility that with the advertisement of settlement, region, and country, a gastronomic festival could participate in marketing projects, tourism development or rural development projects.

People of any rural region, with the help of their own gastronomic festival, can introduce their own products, food and drinks. It is a good chance, for example, to show the products of bio-cultivation, which although they probably cost more, will be bought for their health benefits.

In Hungary, an association has recently been founded: the Association of Hungarian Gastronomic Festivals, which already has 30 members. In 2003, 3 million visitors were attracted to the gastronomic festivals which were organized by the Association.

The aims of the Association are to introduce Hungarian national cuisine by means of gastronomic festivals, in order to popularize them both in Hungary and in other countries. It is also interested in looking at the various Hungarian traditions of culinary culture throughout the regions of the country: there are different ways of food preparation in different regions and different recipes too. The Association plans to collect the many different recipes. Its aim is to propagate the national cuisine in Hungary and abroad. And, in addition, its task is to explore the historical, ethnical, sociological, gastronomical and touristic traditions of Hungarian culinary culture by region, and, last but not least, to explore and keep gastronomic traditions alive for future generations.

From April to October, the programme managers offer many excellent gastronomic festivals. These festivals display a range of culinary delights: fish, sausage, goulash, stewed mutton, beef, meals with paprika (paprikash), food from the traditional slaughter of the pigs, tripe, meals with the meat of wild animals, bread in a traditional baker's oven, edible flowers, etc.

About 262 gastronomic festivals were held in Hungary in 2004. If we take a look at the distribution of gastronomic festivals in Hungary by month (Figure 6.1), we will see that from May to September the numbers are really high. On 1 May (a national holiday), there were many gastronomic events and cooking competitions, and in September there were many grape and wine festivals. In the summer, August is the busiest month, but interestingly, in July, less gastronomic festivals take place than in May or June.

In Hungary, 35 per cent of the gastronomic festivals also include some form of cooking or wine competition (see Figure 6.2). The most popular gastronomic events

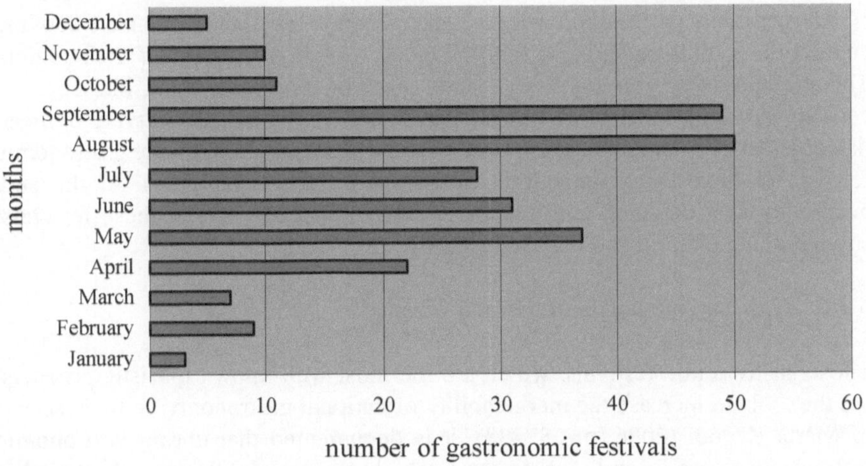

Figure 6.1 The distribution of gastronomic festivals in Hungary by month

Source: Own data collection.

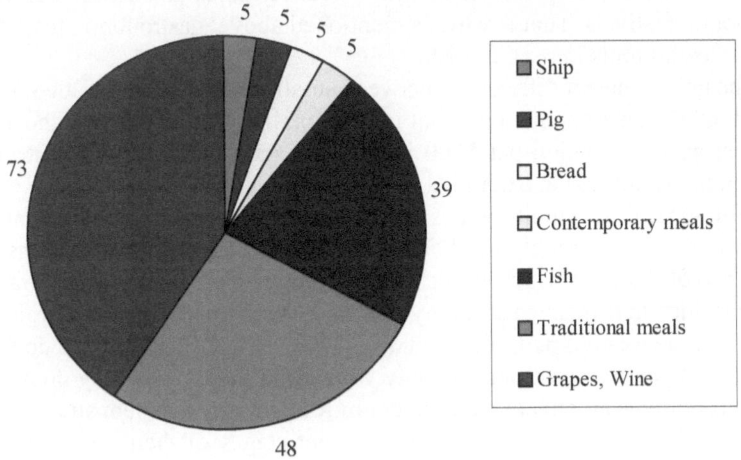

Figure 6.2 The most celebrated food products of Hungary (as a percentage of consumption)

Source: Own data collection.

in Hungary are wine festivals, grape harvests, wine competitions and exhibitions. There are also gastronomic festivals that are variations on the traditional meals of goulash, paprikash, sausages, fish and fish-soup; a few festivals which try to recapture the food of medieval times; and festivals based on mutton and pork; and last, but not least, since it is much appreciated by Hungarians: bread.

During many gastronomic festivals, people cook together or/and have cooking competitions, thus celebrating food or meals which are typical of their region. This explains why you can find various types of traditional food, like open-air cooking in big pots: fish soup, gulash, paprikásh, etc. with different types of meat, like chicken, rooster, lamb, pork, beef, mutton, game, all in the same gastronomic festival. If there is a competition, there are different categories. At the end, everybody eats together, and that is the most important part of these festivals: strengthening cultural and regional identity.

6.3.2 Gastronomic potential of small regions

Those gastronomic festivals, which are the most well-known touristic products, are the tools to increase the marketability of national gastronomy. In the research of Márta Kóródi (2006, pp. 87–88), it is documented that during gastronomic festivals local people and tourists get much closer to each other than in the other forms of tourism.

These events capture the traditional atmosphere, with traditional instruments, traditional installations, and most of the time there are some additional programmes such as folk dances, folk music, traditional conventions, and artisans during the gastronomic festivals. That is why, as mentioned above, gastronomic festivals are the most useful tools in rural tourism.

To be able to detect differences between small regions, Kóródi (2006, pp. 89–90) defined the gastronomic potential for the small region as follows: the amount of tourist accommodation per 1000 inhabitants and the number of gastronomic events in the small region examined.

In Hungary, those small regions which have a high gastronomic potential are those which have a strongly rural character. There are some small regions in the eastern part of Hungary, which do not have anything else but gastronomic attractions. They consider that their gastronomy can be a leading tool in the development projects. In the western part of the country, however, the small regions do not pay too much attention to gastronomy (only wine), and that is why they do not think that gastronomy is an effective ingredient of tourism-development strategies.

This difference occurs because of different stages of their development. In Hungary (Bernát and Zoltán, 2003), some complex rural tourism services could provide the impetus for development in disadvantaged zones.

6.4 Fish soup festival in Baja

Hungary has two big, main rivers: one is the Danube and the other is the Tisza. There are also many important lakes including Lake Balaton and Lake Velencei. These rivers and lakes are rich in fish. In the past, people who lived near the lakes and rivers made a living out of fishing. They worked essentially with the same methods, but every area had its own typical fish soup which marked the

Table 6.2 Fish soup festival in Baja

	Years of Baja's Fish Soup Festival									
	1996	**1997**	**1998**	**1999**	**2000**	**2001**	**2002**	**2003**	**2004**	**2005**
Fish (t)	3	4,5	5,5	7,8	12	15	15	15	15	15
Onion (100kg)	3	4	5	8	11	13	13	13	13	13
Red paprika (kg)	150	200	250	320	500	520	550	550	550	550
Water (l)	4000	5200	6400	8500	13000	13000	14000	14000	14000	14000
Baja's fish soup pasta (100kg)	2	3	4	6	10	14	14	14	14	14
Registered space for cooking	463	687	948	1300	2014	2014	2014	2014	2014	2014
Number of visitors	8000	15000	20000	25000	45000	60000	70000	72000	75000	70000
Number of organizers	150	150	150	250	500	600	650	700	700	700
Costs of event (in thousand HUF)	3800	5800	7600	9000	12000	18000	20000	22000	25000	27000

Source: Data on Baja's Fish-Soup Festival (Bajai Népünnepély Journal (2001), and own collection).

differences between regions. Nowadays, the culture has conserved its gastronomic tradition, and in summer every area has its own fish-soup festival in order to recall old times and traditions, to put people in touch with each other, and to appreciate nature's fine harvest.

The town of Baja with all its natural resources is a special tourist attraction. The city has been home to 40,000 inhabitants for many centuries and Hungarians, Germans, Croatians and Serbians have peacefully cohabited. This cultural diversity will always be preserved through the local buildings and traditions. The Danube and its tributary, the River Sugovica, which crosses the town centre, attracts water-sports fans and fishermen.

Each year the famous Baja fish soup/fishermen's soup and the wines specific to the region are enjoyed by thousands of people during the Baja Fish Soup Festival

This famous fish-soup cooking event at Baja was mentioned in the Guinness Book of Records, with more than 2000 boiling pots. Today, Baja proudly bears the title of 'Fish-soup Capital', with a dish that has become a symbol of Baja.

In the article entitled 'Something fishy in Hungary? It's not all goulash' about the Fish-Soup Festival and the fish-soup, Roddy (2005), commented: 'Hungary is famed throughout the world for its fiery, meaty goulash, but this nation of food fanatics has a lot of other arrows in its quiver, one of which is its hearty fish soup.' Baja celebrated the 300th anniversary of becoming town in 1996. To mark this occasion the town organized lot of programmes, but there was a special idea to organize one big common eating event, of course, just with a traditional meal, known all over Hungary, Baja's fish soup with pasta. The organizers held it at the most convenient place: the main square next to a tributary of the Danube. The motto of this programme was: '300 years old town, 300 kettles of fisherman's soup'. The organizers attracted great interest, so instead of 300 kettles there were 463 kettles in the first year. For every kettle, there was one table where everybody could welcome ten or more guests. Table 6.2 now shows the growth and the stagnation of the festival.

In 2000 with 2014 boiling kettles the festival reached its limits. There were no more places in the heart of the city to set up more kettles, and the organizers had to secure the place (ambulance, fire-fighters...), and rethink the main objective of festival.

When looking at the impacts of the festival, it can be argued that the rise of more than 60,000 people passing through the city imposes a heavy environmental burden, although it was a massive economic hit for the service sector and a great deal. The most important effect is the social one, because year after year people go back to enjoy the event. Some of them plan vacations there in the second weekend of July. This event is a great example of a gastronomic festival based on an ancient tradition that had been revived to remember old times, by eating and being together.

6.4 Reflections for the future

In the new European development, stronger emphasis is being placed on regional development and regional identity. For many regions, the productions of food and gastronomic traditions and events have become important constituents of regional identity.

The base is set: we have saved our own cultural traditions in gastronomy and there is an interest from tourists, too. I am convinced the concept of that culinary adventure could help in rural development as well as in sustainable development. This development needs well-thought marketing projects and the cooperation of people working at the national and regional levels. Hungarian wine harvests, wine trails, gastronomic roads and gastronomic festivals have their roots in old traditions. This fact enables us to rethink the role of gastronomy in our life. In this globalized world, traditional food and drinks have become tourist attractions.

For the future we have the opportunity to build one Disney-World-type, huge gastronomic adventure park to be called the 'Gastro Amusement Park' or 'GASTROLAND', that would straddle the Danube north of Budapest, taking in three Hungarian towns and one Slovakian, offering visitors a chance to sample cuisines from dozens of nations. So far, 43 countries have indicated that they are interested in joining the project, which will start in 2008.

According to hirado.hu and vendegvaro.hu, 'Gastroland' would extend over the area around Szob, Ipolyalmásd, Márianosztra (Hungary) and Chlaba (in Slovakia), and would present the culinary delights of the world, taking advantage of the area's diverse topography. For example, the representatives of Mexico will open their restaurant in the mountains, the Dutch near water, the Austrians near pine forests, and the Hungarians will build a 'csárda' (typical Hungarian restaurant) on some low-lying ground. There are plans for an international gastronomical museum and a total of 50 catering trade units, each with the ability to hold 200–300 diners at a time. The project will create an estimated 4,000 jobs (http://www.pestiside.hu).

The aim of this chapter was to give a general overview of Hungary's gastronomic cultural heritage and reserves. I assume our traditions are strong enough to build rural development strategies on them using elements of rural tourism and marketing projects. The sustainability of rural development is substantially supported by deep cultural roots which also serve to ensure continuity.

References

Bajai Honpolgár (2002), Várostörténet és Környezetvédő lap, XIII. évfolyam 7. (141.) szám, 2002 július

Bajai Népünnepély (2001), 'Alföldi Lapok', Alkalmi Kiadvány, július 14.

Bernát, T. and Zoltán, Z. (2003), 'Hátrányos helyzetű mezőgazdasági körzetek és piacgazdaság', AGROINFORM Kiadó és Nyomda Kft., Budapest, p.152.

Britton, S. (1991), 'Tourism, capital and place: towards a critical geography of tourism', *Environment and Planning, D*: Society and Space, vol. 9, pp. 451–78.

Butler, R. and Hall, M. (1998), 'Conclusion: The Sustainability of Tourism and Recreation in Rural Areas', in: R. Butler, M. Hall, and J. Jenkins, (eds), *Tourism and Recreation in Rural Areas*, Toronto, Wiley, pp. 249–258.

Bywater, M. (1993), 'The market for cultural tourism in Europe', *EIU Travel and Tourism Analyst*, vol. 6, pp. 30–46.

Craik, J. (1995), 'Is cultural tourism viable?', *Smarts*, vol. 2, pp. 6–7.

Crompton, J. L. (1979), 'Motivations for Pleasure Vacation', *Annals of Tourism Research*, vol. 6, pp. 408–424.

Crompton, J. L. and McKay, S. L. (1997), 'Motives of Visitors Attending Festivals', *Events Annals of Tourism Research*, vol. 24, no. 2, Elsevier Science Ltd, pp. 425–439.

Dower, M. (1998), 'The Heritage as a Resource', LEADER Magazine, no.17–Spring.

Fleischer, A. and Felsenstein, D. (2000), 'Support for Rural Tourism: Does it Make a Difference?', *Annals of Tourism Research,* vol. 27, pp. 1007–1024.

Fredericks, M. (1993), 'Rural Tourism and Economic Development'. *Economic Development Quarterly*, vol. 7, pp. 215–226.

Furi-Perry, U. (2005), 'Not Just Known for Goulash: A Tour of Hungarian Wines', 1/4/2005 available at http://www.buzzle.com/editorials/1-3-2005-63699.asp.

Gulyás Hírmondó (2002), 'IV. Szolnoki Gulyásfesztivál', Szolnok, szeptember 7.

Halászlé Ünnep (2004), 'Bajai Peőfi Népe Különkiadvány', július 7.

Halászlé Ünnep (2005), 'Bajai Peőfi Népe Különkiadvány', július 6.

Handler, R. and Linnekin, J. (1984), 'Tradition, genuine or spurious', *Journal of American Folklore*, vol. 97, pp. 273–90.

Hill, B. (1993), 'The Future of Rural Tourism', Department of Parks and Recreation, Washington DC, at http://www.ersa.org/ersaconfs/ersa04/PDF/314.pdf (application/pdf)

Horváth, L. (ed.) (2004), 'Paprikás ételek bográcsban', Kecskemét, Béta Press Stúdió Kft., p.96.

Iso-Ahola, S. E. (1980), *The Social Psychology of Leisure and Recreation*, Dubuque IA, Wm. C. Brown.

Jolliffe, L. and Baum, T. (1999) 'An Agenda for Cultural Tourism on the Periphery: The Case of Four North Atlantic Islands', Paper presented at CHME Conference, University of Surrey, Guildford.

Kochel, U. (1994), *Culture, Tourism, and Development: The Case of Ireland*, Liverpool, Liverpool University Press.

Kóródi, M. (2006), 'A vidék- és a vidéki turizmusfejlesztés összefüggései a magyarországi kistérségekben', Draft of Ph.D. thesis, Szent István University, p.118.

Kovács, D. (2003a), '*Falusi vendéglátás Magyarországon*', Agroinform Kiadóház, Budapest, p.123.

Kovács, D. (ed.) (2003b), '*A falusi turizmus hagyományai*', Mezőgazda Kiadó, p.224.

Kovács, F. and Kovács, J. (eds), (1999), 'A falu- és vidékfejlesztés stratégiai kérdései', MTA Agrártudományok Osztálya, Agroinform Kiadó, Budapest, p.100.

Lewis, J. (1998), 'A Rural Tourism Development Model', *Tourism Analysis* vol. 2, pp. 91–105.

Lundberg, D.E., Krishnamoorthy, M. and Stavenga, M. H. (1995), *Tourism Economics*, ISBN: 978-0-471-57884-0.

Mansfeld, Y. (1992), 'From Motivation to Actual Travel', *Annals of Tourism Research* vol. 19, pp. 399–419.

Millar, S. (1989), 'Heritage management for heritage tourism', *Tourism Management*, vol. 10, n° 3, pp. 9–14.

Narvud, S. and Ready, R. C. (2002), *Valuing Cultural Heritage: Applying Environmental Valuation Techniques to Historic Buildings, Monuments and Artifacts*, Cheltenham, UK, Edward Elgar.

Nosi, C. and Zanni, L. (2004), 'Moving from "typical products" to "food-related services": The Slow Food case as a new business paradigm', *British Food Journal*, vol.106, n°s 10–11, pp.779–792.

Ohlsson, T. (2000), 'Regional Culinary Heritage, A European network for regional Foods', Paper presented at the International Conference on 'Local Food and Tourism', 9 -11 November, Cyprus Tourism Organization, Cyprus, with the support of the World Tourism Organization, at http://www.culinary-heritage.com

Pearce, P. L. (1982), *The Social Psychology of Tourist Behavior*, Oxford, Pergamon Press.

Prentice, M. (1993), *Tourism and Heritage Attractions*, Routledge, London.

Prohaska, S. (1995), Trends in Cultural Tourism, in: M. Conlin and T. Baum (eds), *Island Tourism: Management Principles and Practice*, London, Wiley, pp. 33–51.

Riganti, P. and Nijkamp, P. (2004), 'Valuing cultural heritage benefits to urban and regional development', Paper presented at the 44th European Regional Science Association Conference, University of Porto, 25–29 August 2004, Porto.

Roddy, M. (2005), 'Something fishy in Hungary? It's not all goulash', Story from REDORBIT NEWS: http://www.redorbit.com/news/display/?id=179424, Published: 2005/07/19 07:47:34 CDT © RedOrbit 2005.

Rojek, C. and Urry, J. (eds) (1997), '*Touring Cultures – Transformations of Travel and Theory*', Routledge.

Santich, B. (2004), 'The study of gastronomy and its relevance to hospitality education and training', *International Journal of Hospitality Management*, March.

Schlüter, R. (2002), 'History, Culture and Identity, The Immigrants' Heritage in South America Food and Culture as a New Sustainable Tourism Product', Tourism in the Americas, *Development, Culture and Identity*, Winter 2002, at http://drclas.fas.harvard.edu/publications/revista/Tourism/Schluter.html

Sharpley, R., Sharpley, J. and Page, S. (1997), *Rural Tourism: An Introduction*, London, International Thomson Publishing.

Smith, M. K. (2003), 'Issues in Cultural Tourism Studies', Routlege.

Tassell, C. and Tassell, M. (1990), 'The Tasmanian Rural Landscape', *Heritage Australia*, vol.9, n° 4, pp. 12–15.

Timothy, D. J. and Boyd, S. W. (2003), *Heritage Tourism, Pearson Education*, p. 327.

Tusor, A. and Sahin-Tóth, Gy. (1999), '*Gasztronómia – Étel- és Italismeret*', Kereskedelmi és Idegenforgalmi Továbbképző Kft. Budapest.

Uysal, M., Gahan, L. and Martin, B. (1993), 'An Examination of Event Motivations: A Case Study', *Festival Management and Event Tourism*, vol. 1, pp. 5–10.

Weiler, B. and Hall, C. (1992), *Special Interest Tourism*, Belhaven, London.

Whitehead, J. C. (2003), Reviewed for EH.NET, Valuing Cultural Heritage: Applying Environmental Valuation Techniques to Historic Buildings, Monuments and Artifacts.

Williams, R. (1958), 'Culture is ordinary', in: R. Gale (ed.) (1989), Resources of Hope – Raymond Williams, London, Verso.

Wolf, E. (2002), 'Culinary Tourism: A Tasty Economic Proposition', Available from http://www.culinarytourism.org/resources.htm> (accessed 2 September 2002)

WTO (1994), *National and Regional Tourism Planning, Methodologies and Case Studies*, London, Routledge.

Websites

http://www.arts.adelaide.edu.au/centrefooddrink/articles/howmuchdrink.html
http://www.borut.hu/_eng/koszonto.htm
http://www.capvex.com/heritage.shtml
http://www.cheflaszlo.com/traditionalhungarian.html
http://www.hirado.hu
http://www.hungary.com
http://www.mmm.fi/maasmatk/lehti/03sRural_Tourism.htm
http://www.pestiside.hu
http://www.vati.hu
http://www.vendegvaro.hu

PART II
Traditional Markets and Globalization

Chapter 7

Meat Processing in Ibero-American Countries: A Historical View

Javier Mateo, Irma Caro, Ana Cristina Figueira, Daphne Ramos and José M. Zumalacárregui

Introduction

At present, humans are omnivorous and probably our early hominid ancestors were, too. Food, with a basic function of keeping us alive and healthy, is an integral part of the human experience and its history, for the most part, is based on archaeological materials – and thus on a number of assumptions. The major techniques employed for food preservation have long since been used. People from prehistoric cultures have left evidence of their knowledge of many primitive technological operations, such as drying, preserving moist foods with brine or salt, freezing, fermenting, heating, etc. (Peterson, 1963). Food processing must have been developed gradually, with human beings exploiting the earth's resources more efficiently by mastering manual and technological skills, and developing social organization by the time that civilized communities were founded (Blucher, 1991).

Since antiquity, human nutritional patterns have depended on available environmental resources and cultural factors, marking the differences between regions. Societies have been continually changing, and social changes have been reflected in food, eating, and nutritional patterns (Sobal, 1999). As a result of historical human movements, conquests and trading, today we have reached a point where the history of many traditional meals is actually based on the influences of several former cultures, from different regions of the world.

One of the most dramatic developments in agricultural and food processing in the last millennium occurred in the New World, after the arrival of the European colonizers, and had worldwide consequences. Amongst the consequences of those changes, the most remarkable is the increase in meat consumption in America (Blucher, 1991).

This study deals with the Latin American traditional meat products, from a historical perspective. As a matter of introduction, throughout the study, the authors would like to draw the readers' special attention to the resemblances and differences between the meat products of Latin America and those of Europe, in particular those of the Iberian Peninsula. Simultaneously, we would like to encourage the readers to reflect on whether the common knowledge of Ibero-

American meat processing may represent opportunities for innovative actions in peripheral areas.

The main parts of the present work are listed below.

- One provides a historic perspective of meat processing in Latin America.
- The other deals with existing Latin American meat products.

With respect to this last part, typical Latin American meat products have been grouped in three main categories, according to their origin:

- 'Charqui' and similar sun or wind dried meat products – dried outside or in open spaces;
- Creole meat products;
- European-like meat products.

The first group is comprised of salted and partially dehydrated products made from very thin slices of meat, which are cooked before consumption. They are believed to originate from the ancestral 'charqui', which was already known in America in Pre-Columbian times. These products can be considered as Intermediate Moisture Foods, with water activity (aw) values between 0.60 and 0.90.

Creole meat products could be defined as traditional Latin American products which originated from the adaptation of former Iberian meat products that were brought to America by the colonizers (usually Iberians). Creole products would have developed from those, as a result of the adaptation to local environment and resources, extended contact between communities, mixing bloods and blending cultures.

The third group has the same – or almost the same – characteristics as the actual European products, which are more important in populations where direct descendants from Europeans are abundant: for instance, throughout most of Argentina and Chile, in parts of Brazil, etc.

7. 1 Historical perspective

Meat consumption must have been a milestone in human evolution. Eating meat compared with eating vegetables implies a longer time for digestion between-meals, which would allow the prehistoric human being to spend more time carrying out activities different from gathering fruits and vegetables to be eaten. Hunting and animal breeding skills were developed in this way.

It seems logical to think that the advantages of saving possible meat surpluses encouraged humans to search for meat preservation methods. Empirical meat preservation technologies based on salting, fermenting, drying and smoking were thus developed (Zeuthen, 1995). For example, evidence for meat salting and sun-drying by the ancient Egyptians has been reported (Pearson and Tauber, 1984).

Sausages, consisting of a mixture of small-sized or minced meat, salt, spices and other ingredients, which was put in a casing, must have been a rather sophisticated invention. The earliest proof of the oldest sausages dates back for more than 5,000 years and was found in the Sumerian culture – the former Babylonia. The first written references about sausages date from approximately 600 BC, both in ancient China and in Greece. In China a sausage known as 'Lachang' – sweetened, seasoned and smoked – was first mentioned, and in Greece sausages were described in the works of Homer and Epicharmus, the latter having written a comedy entitled 'The sausage' (Anon, 2006).

Swine was one of the principal domesticated livestock species used for meat production. The first swine farms would probably have appeared by 10,000 BC in the near East (Delgado et al., 2004). Accompanying human migrations, the presence of domestic swine was expanded to other parts of Asia, and also to Central-European and Mediterranean areas. Thus, three swine types – respectively Asian, Celtic and Mediterranean – were differentiated, from which, by means of different cross-breeding, the present swine breeds originated.

Pig breeding was importantly developed in ancient China, Egypt and Greece. The Ancient Greeks and Romans used pork extensively for meat product manufacture (Anon, 2006). Subsequently, the use of pork has prevailed from the Middle Ages and the Renaissance to present times in the Christian world. However, pork consumption has been an element that distinguishes Christians from both Jews and Muslims throughout the history of the Abrahamic religions.

Traditions and techniques developed by the Ancient Greeks were shared with, and improved by, the Romans. Meat products became very popular and favourite dishes for the Romans, who mastered the art of curing meat. The Romans were probably the first to note the reddening effect now attributed to nitrite (Ray, 1994) – nitrate would be present originally as a natural impurity in salt used for curing and, unknown to the users, became a key ingredient in the curing process. Meat for drying was salted and then often brushed with oil and then covered in a mixture of dried herbs, crushed pepper, or ashes before being hung high up either in the smokehouse or in a cool, dry place (Shepard, 2000).

Contemporary Roman texts described a range of typical meat products of the times: *botulus*, *lucanica*, *farcimen mirtatus*, *pendulus*, *hilla*, *tomacula*, *circeli*, *perexsuccus*, etc. (Giacosa, 1992; Aguirrezábal, 1993; Shepard, 2000). Some of those names may remind us of current names of Italian, Portuguese or Spanish traditional meat products. For example, it is possible that the Latin term 'perexsuccus', which meant very dry, was used for a pork ham salted and long-ripened, and the actual Italian term 'prosciutto' is derived from this (Personal communication).

Meat products were also linked to the social aspects of Roman life. For instance, the law punished manufacturers whose meat products had been the cause of food poisoning; and sausages, known by the Latin word '*farcimen*', were ritually and symbolically present in several recreational feasts, such as the Lupercalia festival – which is said to be the origin of the current St. Valentine's day (Coxe and Coxe,

1987; Panati, 1989). In this sense, the early Catholic Church opposed the pagan and libertine Roman festivals and banned sausage consumption during its first years of existence. Blood consumption, as a religious matter, was banned by the Council held in Nicea (325 AC) and was banned again by the Byzantine Emperor Leo VI, 'The Wise' (ruled 886–912), on the grounds not only of religion, but also of public health, probably as an empirical preventive measure against outbreaks of botulism (Erbguth, 2004).

However, in spite of prohibitions, sausage production and consumption were common activities during the Middle Ages in the Christian Europe. Later on, during the Renaissance, although based on former traditions and carried out by small manufacturers, the art of making meat products experienced a new development. In the 15th and 16th centuries, the production of salted, cured and dried meat products grew to supply the men at war or on expeditions travelling with food, and those meat products also formed a part of the scanty diet of the country people and the poor (Wright, 1999). In the colder regions of Europe, where a cold climate makes it possible to preserve perishable food for quite a long time, fresh, raw semi-dry, and cured-smoked-cooked meat products were mainly manufactured. Meanwhile, in the warmer South, the main meat products consisted of raw dry-ripened meats, sometimes fermented, which were dried in the coldest months of the year and thus preserved for the hottest ones. Many kinds of European sausages were justly famous; farming families would make their own sausages and sell them in town. Muslim people also knew how to make salted meat and sausages, made of lamb or beef (Wright, 1999), such as the Turkish dried beef 'pastirme' (the root of the word *pastrami*).

In this context, people from Iberia arrived in Latin America and brought with them livestock, i.e. cows, horses, sheep, pigs, and their meat product 'know-how'. Regarding swine, Mediterranean local breeds, i.e. 'Black Mallorquín', 'Alentejano', 'Spanish Iberian', 'Black Canario' and Celtic ones, i.e. 'Celtic Gallego', 'Portuguese Bísaro', were introduced into America, and it is from these that the Latin America Creole pigs and local breeds have originated (Delgado et al., 2004).

The Americans had preserved meat prior to the Europeans' settlement, by drying and eventually smoking hanging strips or pieces of meat (Ray, 1994), and had probably developed products such as 'charqui' (thin dried strips of meat, a kind of jerked meat) or pemmican (powdered air-dried buffalo meat which was pounded together with melted fat, dried fruits and vegetables) (Pearson and Tauber, 1984). Nevertheless, the Iberian 'know how' about meat products was steadily instilled in the Latin American territories and cultures, but techniques had to be adapted to cope with the local climate, resources, culture, and later on with the cultural influence of other European, African and Asian immigrants. The descendants of settlers would try to assimilate their inherited knowledge by carrying out modifications whenever they came across specific different circumstances, in order to both improve and adapt processes and products (Monín, 1991). On the other hand, changes in traditional meat processing in Iberia originated from Latin

American influences. It is worth mentioning the use of *Capsicum annuum* plants – paprika – for meat-product manufacturing in Iberia, and, usually to a lesser extent, in other parts of Europe (Zapata et al., 1992).

Accordingly, with the passing of time, traditional meat product processing in Latin America has become diverse in many ways. Latin America is a huge region in the world – much larger than the Iberian Peninsula. Its physical geography is complex with plenty of contrasts: mountain ranges all along the West, high plateaus, forests, valleys, deserts and coast. The climate is no less variable, including arid and semi-arid, dry and wet-tropical and subtropical zones, warm areas, cold mountain climates and even frozen and polar zones (Mata et al., 2001). Diversity is multiplied by cultural and social factors, mainly based on racial and ethnic contacts and a large economic gap between the rich and the poor, which has unleashed a deterioration of peripheral areas and strong migration movements to large cities giving rise to urban poverty belts (Mata et al., 2001). All this variability has been reflected in the diversity of Latin America's agricultural, livestock production and agro-industrial activities. To sum up, there is a considerable variety of traditional meat products in Latin America. The main typical meat products in Latin America can be classified as follows according to their historical origin:

- A group of intermediate moisture meat products formed by the ancestral charqui and other related meat products, which are made from thin pieces of non-minced meat.
- Creole meat products, which could be defined as meat products which were first introduced in Latin America some centuries ago by immigrant colonizers, but since then have evolved into American traditional varieties.
- A set of meat products, practically the same as their European counterparts.

Evidence of the common past of Iberian and Latin American traditional meat products can be observed not only for the latter group of meat products. It is possible to find some charqui-derived or Creole meat products from Latin America with the same name as some Iberian traditional meat products, but showing different appearance or flavour according to the particularities of each region. Some examples are listed below: i) a variety of fresh sausages called 'Chorizo/ Salchicha/Longaniza' or 'Chouriço/Salsichão/Linguiças' (Spanish- or Portuguese-speaking countries, respectively); ii) a range of types of 'Chicharrones', which are mainly defined as the solid remains of lard, skin and meat after the liquid fat has been removed by melting and straining it; iii) a collection of partially dehydrated meat products called 'cecina' with differing ingredients in their formulation, shapes, sizes and degree of dehydration. Conversely, it is also possible to find meat products with similar appearance or flavour but with different names.

However, by comparing traditional Iberian and Latin American meat products, the following facts can be highlighted. On the one hand, the consumption of meat products intended to be eaten raw: namely, dry-ripened sausages or ham, is not so popular in Latin America as in Iberia. On the other hand, salted partially-dehydrated meat products, to be cooked before being eaten, are more popular in Latin America than in Iberia. Finally, a greater percentage of beef is produced and processed in Latin America than in Iberia (FAO, 2006; http://faostat.fao.org/).

Both in Europe and Latin America, although meat products are produced more and more on an industrial scale, typical handmade meat products have managed to maintain a market niche. Looking to the future, a number of innovative trends in meat product processing and consumer expectations have been reported in several reviews (Issanchou, 1996; Resurreccion, 2003; Arihara, 2006; Leroy et al., 2006). Innovation does not seem to be incompatible with small-scale enterprises – such as those of peripheral areas producing traditional meat products – in which the use of appropriate (i.e. ecological, local, functional, high intrinsic quality, novel, etc.) ingredients or raw materials and the application of low cost and affordable new technologies would be key factors in developing innovative actions.

7. 2 Latin American meat products[1]

7.2.1 'Charqui' (or charque) and other related dried meat products

In Quechuan the word for 'charqui' (or charque) is '*ch'arki*', and the origin of this meat product has been lost in history. In the Andean region, since pre-Columbian times, charqui has been manufactured with llama meat and game (LeMaguer and Jelen, 1986), and at present beef and horse meat is also used. Andean charque is usually made with lean meat that is cut into strips, and salted and dried outside with the help of the wind and the sun. Once the product is dry enough, a second particle size reduction is carried out before it is cooked. According to Collazos et al. (1993), the approximate composition of beef charqui is: 26 per cent moisture, 58 per cent protein, 4 per cent fat and 12 per cent ash. Charque is used as an ingredient in several traditional recipes.

Charque and other air-dried intermediate-moisture meat products (IMMP) – products with a water activity a_w between 0.60 and 0.90 (Leistner and Rodel, 1976) – are widespread in Latin America. The manufacturing processes of traditional Latin American dried meat products present multiple variations due to meat animal species, cut-size of meat, salting conditions, addition of spices or condiments, drying conditions (time, temperature – up to 70 °C, air humidity, outside or indoors, smoking), and post-drying processing. All these variables result in differences in appearance, texture, flavour and functionality of different products.

1 The reader can see photographs of Latin American meat products mentioned in the text on the following authors' web page http://www3.unileon.es/personal/wwdhtjmo/photos.doc.

One other Andean traditional IMMP product similar to charque is 'Chalona' which consists of pieces of sheep meat on the bone, which are salted and then dried. Another one is 'Cecina' made from beef or pork steaks, which are salted and briefly dried by the wind, under the sun, near cookers, red-hot coals or bonfires, at different temperature and humidity, according to the local weather and drying system. 'Cecina' contains more moisture and much less salt than charque or 'chalona'. It is generally made from beef or pork In the case of pork 'cecina', the surface of the meat is impregnated with a paste made with salt, liquid fat, Capsicum spp, annatto, spices, lemon juice, etc. The water activity (a_w) of pork cecina is between 0.90 and 0.98 (unpublished data).

Finally, one more traditional product known as 'carne seca' (the Spanish for dry meat) is mentioned. This product consist of thin pieces of lean meat which are salted, dried with hot air (c.a. 70 °C) and then cooled and shredded or ravelled into threads.

In the Pampa, a specific traditional 'Charque' is produced. Industrial production of Pampean 'Charque' goes back to the 18th century (Monín, 1991) and had an important economic and social impact until the first decades of the 20th century. That 'Charque' was exported to England and other parts of the world (Camargo et al., 1984). However, at present, Charque is only produced as a traditional meat product and is commonly used as an ingredient of 'feijoada' a traditional dish.

Traditional Pampean charque is made with flat deboned beef pieces – with a high surface area and relatively low uniform thickness, up to 5 cm. Meat pieces are submitted to a dry salting process by staking meat into piles separated by layers of coarse salt. Meat is periodically restacked so that the upper meat pieces are positioned at the bottom, and vice versa. After about 3–5 days of salting, the pieces of meat are washed to remove excess salt from the meat surfaces and then sun dried, and in the evening or on cloudy days meat pieces are collected, piled and covered in a roofed space. After enough drying time (5–15 days), the charque is packed in bales, placed inside a sack and stored; refrigeration is not needed. In more recent times, charque has been packaged in retail sized vacuum packs, giving it an even longer shelf-life and a much improved appearance (Norman and Corte, 1985; Torres et al., 1994). According to the parameters of the Brazilian official inspection, traditional charque should contain up to 50 per cent moisture and 20 per cent salt and less than 0.75 final values for a_w; fat content is normally around 6–10 per cent (Torres et al., 1994; Pardi et al., 1996).

A derivative of charque popularly known as Jerked Beef has recently been introduced into the market, which differs from the former by the use of curing agents, both dry and wet salting, and vacuum packaging (Pardi, 1996; Pinto et al., 2002). Wet salting by the immersion of beef pieces may take up to 24 hours – pumping brine and massaging being optional – while the dry salting lasts for up to 72; weight losses after washing should be about 20 per cent and drying takes place at 25–35 °C (Norman and Corte, 1985). Jerked beef is officially characterized by the Brazilian legislation as having a maximum moisture content of 55 per cent, 50 ppm sodium nitrite, 18 per cent ash, and a final a_w lower than 0.78 (Youssef et al., 2003).

In the 'Nordeste' region of Brazil a type of 'charque' made with beef or goat meat known as 'carne-de-sol' (sun meat) is produced. Because of differences in weather conditions, it is dried for a shorter time (up to 3 days) and at higher temperatures. 'Carne-de-sol' has a stronger flavour and is darker than Pampean charque. It has both higher moisture and lower salt content and thus it has a limited shelf life. The product is characterized by salt levels of between 5 and 6 per cent and moisture levels between 64 and 70 per cent, rendering a range of a_w between 0.91 to 0.95 (Norman and Corte, 1985).

Besides charqui, in Argentina and Uruguay, a traditional salted and dried meat called 'Tasajo' is also produced. The meat pieces for 'tasajo' are considerably smaller and thinner than for charqui (c.a. 2 cm thick strips) and drying is carried out in brine for less than an hour. Afterwards, the meat pieces are dried in a similar way to charqui (Zumalacárregui and Díez, 2001).

More to the North of America, in México, there is a variety of traditional IMMP, known as 'cecina', which is made from either beef or pork, the latter impregnated with chilli and other spices. The appearance and texture of 'cecina' is similar to fresh meat but it has a darker colour and is slightly drier (Reyes-Cano et al., 1994). The manufacturing method for Mexican beef cecina includes cutting the meat into long slices (2–8 mm thick, up to 1 m long, 10–25 cm wide) in the direction of the muscle fibres, a salting step, with the addition in some cases of oil and vinegar, and the drying of salted meat, hung up under the sun's rays or in ovens at 50 °C for a variable length of time, e.g. 4–24 hours. 'Cecinas' so prepared are folded and covered with paper for distribution. Beef 'cecina' is an IMMP (moisture 61–65 per cent, salt content 8–10 per cent, and a_w, 0.85–0.90). The shelf life of Mexican 'cecina' is some days at room temperature, and can be considerably extended at temperatures lower than 10 °C.

There is another traditional Mexican dried meat product known as 'Machaca', which is similar to the above-mentioned Peruvian 'dried meat'; and in fact it is also a hot-air dried meat product. The key operations in 'Machaca' manufacturing are cutting the meat into slices, dry-salting until 2–7 per cent salt is reached in the meat, drying, first in the sun and then with the help of heated air, and crushing or shredding the salted dried meat with a mortar or a mill to render fibre-shaped particles (González-Méndez et al., 2002). 'Machaca' is usually consumed after cooking it with eggs, onions and chile peppers.

Apart from Latin American IMMPs, other IMMPs are traditional in several parts of the world, e.g. jerky in the USA or Australia, dendang in Southeast Asia, bitong in South African, kilishi in Nigeria, and pork jerky in China.

7.2.2 Creole meat products

Creole meat products could be defined as traditional Latin American products which were originated from the adaptation of former Iberian meat products (brought to America by immigrants) to local conditions and circumstances. This phenomenon involved an innovation process at that time. In this section, meat products originating from cultures other than Iberian are also considered. The majority of Creole meat products consist of both fresh and cooked meat products.

Fresh meat products are defined as those whose meat tissues have not lost the histological structure of fresh meat and spoilage takes place in a few days. Fresh sausages, burgers, marinated meats are examples of this type of meat products. The popularity of fresh meat products in Latin America is high, and is reflected in the countless varieties of these products all over that region.

Fresh sausages form an important group amongst fresh meat products. Fresh sausages are usually made from pork meat and fat, although a proportion of beef, poultry and even meat from exotic animals may also be used. Local varieties of sausages differ from each other in formulation, i.e. animal species, relative proportions of meat and fat, spices and condiments, and in process technology, i.e. meat size reduction, diameter and type of casing, or whether smoking or short-drying processes are carried out or not. In spite of all this variability, almost all fresh sausages are known by one the following names: 'chorizo', 'longaniza', 'salchicha' in Spanish, or 'chouriço', 'linguiça', 'salcichão', in Portuguese.

A considerable number of Latin American fresh meat products can be included in the group of marinated meats, which are usually formed in the shape of steaks, meat pieces for skewers, or larger pieces of meat. Some of these products have been influenced by African and Asian cultures. For example, the origin of Peruvian 'Anticuchos' came from the African slaves, who had no other meat protein dietary source but offal. 'Anticuchos' are made from about 1 inch long pieces of beef heart, and are marinated in a mixture of salt, 'ají panca' and 'ají amarillo' (both *Capsicum* spp), cumin, garlic, pepper, vinegar and oil, and then threaded onto thin wooden sticks. Another example of marinated meat products is 'tacos al pastor' (shepherd's-style tacos). A 'taco' is a traditional Mexican dish comprising a pliable 'tortilla', of either maize or wheat flour, filled with meat, chili-based salsa, onion, cilantro, etc. 'Tacos al pastor' is a Mexican adaptation of grilled meat which resembles the Asian 'kebab', probably brought by immigrants from Lebanon. Although originally made with lamb, most are now made with thinly sliced pork, marinated in herbs and spices, and stacked on a vertical spit, with a narrow end on the bottom and a slice of pineapple at the top. Meat thus prepared is turned in front of a vertical gas flame, shaved off as the outside gets done, and made into 'tacos'. To give one more characteristic example, the Peruvian 'Serrano' smoked ham can be mentioned. It consists of a pork ham without skin in which several deep cuts are made. The scored surface of the ham is then rubbed with a mixture of salt and curing agents, and then kept at chilling temperature for about 3 days. Then it is smoked and a paste made with 'ají panca' and 'ají amarillo' (both *Capsicum* spp), saffron, cumin, 'achiote', oil and garlic is rubbed on the surface of the ham. The ham is then pressed and maintained at a low temperature for a further two weeks. After that, more paste is added to the surface before it is ready to be sent to the consumer (Personal communication). There is a similar 'enchilado' ham in México. For industrially produced-marinated hams, brine is pumped inside.

The very distinctive ingredients of Latin American and Iberian fresh meat products are fresh or dry *Capsicum* spp fruits. Other characteristic ingredients in many parts of Iberia and Latin America are annatto, garlic, vinegar, pepper or cumin.

Amongst the Creole cooked products, one of the most remarkable groups is probably that of blood sausages. Several names are used for blood sausages in Latin America: 'Morcela', 'Morcilla', 'Morcillón', 'Moronga', 'Rellena', 'Prieta', etc. As in the Iberian Peninsula, blood sausages in Latin America have a large regional variability.

7.2.3 European-like products

In Europe, traditional meat products are related to domestic production. Representative examples of traditional meat products of Mediterranean Europe are the dry long-ripened meat products (dry-ripening pork ham and shoulder), dry-ripened beef pieces (Spanish 'Cecina' or Italian 'Bresaola'), dry sausages, etc. On the other hand, cured-cooked-smoked meat products, such as frankfurters, mortadella or cooked ham, are more typical of Central and Northern Europe (Flores, 1997).

In Latin America, a large variety of dry-ripened and cooked meat products practically identical to their European counterparts are produced. These products are traditional in regions characterized by an important presence of European descendants with none or few of mixed-race, where the culture and environmental resources are similar to those of Europe, such as most parts of Argentine, Chile, Uruguay, large parts of Brazil and adjoining areas of other Latin American countries, as well as in large cities around all Latin America. European-like products are often named after the place from which they originally came ('chorizos de Cantimpalo', 'longanizas calabresas', 'morcillas vascas', 'chorizo candelario', etc.).

7.3 Conclusions

In order to arrange the different Latin American traditional meat products in groups, in the present study, a classification according to their origin has been suggested: those intermediate humidity meat products related to the ancestral 'charque'; a bulk of creole meat products resulting from those introduced by immigrants and resembling them in different important features; and European-like meat products.

Most of the Latin American traditional meat products have originated from former European meat products, mainly from the Iberia Peninsula. With the passing of the centuries, meat products have evolved into a huge number of traditional local varieties, which has resulted from not only the time but also the effect of numerous environmental and socio-cultural factors of variability. At present, there is one remarkable difference between Iberian and Latin-American meat processing: in the latter region, meat products are intended to be eaten raw: namely, dry-ripened sausages or ham, are very popular. In contrast, salted partially dehydrated meat products, to be cooked before eating, are much more popular in Latin America.

The manufacturing processes and quality traits of many traditional Latin American meat products have not been scientifically characterized. Instead, products are home-made, empirically, and their processing methods and product characteristics are not conveniently registered. This fact suggests the need for more frequent studies related to traditional meat product standardization and history. The lack of such studies, together with the indiscriminate use of new technologies by small and medium-sized enterprises with the main purpose of reducing costs, presents a serious risk to several Latin American meat products which are in danger of losing their quality and traditional identity. In addition, with the current migration phenomena from rural to urban areas and the subsequent ageing of the population, some traditional meat products known now may disappear, which might result in cultural impoverishment. Furthermore, there is some confusion concerning the names and processing of meat products within and even between Portuguese and Spanish speaking countries. At present, it is possible to find in the Ibero-American community meat products having a common name, but having a different appearance and/or taste; and, vice-versa, having different names but a similar appearance or taste. In this regard, it would be useful to compile a comprehensive list of traditional products, or an Iberian- and Latin American-wide database, including product denomination, definition, historical origin, processing practices and quality standards.

Finally, the common knowledge of the range of traditional products produced by the Ibero-American community might represent a source of inspiration for innovation in the Latin American and European peripheral regions. This might be done in two ways: on the one hand, by means of the adaptation of local meat processing practices to the demands of foreign markets, or immigrant communities. And, on the other hand, by changing the local manufacturing processes, inspired by foreign processing methods, with the aim to expand the available range of meat products in the local markets.

References

Agirrezábal, M.M. (1993), Influencia de diversos ingredientes en el proceso madurativo del chorizo, PhD. Thesis, University of León.

Anonymous (2006), In *Wikipedia, The Free Encyclopedia*, Retrieved March, 2006, at http://en.wikipedia.org/wiki/Sausage#History

Arihara, K. (2006), 'Strategies for designing novel functional meat products', *Meat Science*, vol. 74, pp. 219–229.

Blucher, D.K. (1991), 'History of foods', in: Y. H. Hui (ed.), *Encyclopedia of Food Science and Technology*, Wiley-Interscience Publication, Washington D.C., pp.1404–1412.

Camargo, R. et al. (1984), *Tecnologia dos Produtos Agropecuários*, Nobel. São Paulo.

Collazos, C.,White, P.L., White, H.S. and Viñas, E. (1993), *La composición de los alimentos de mayor consumo en el Perú*, Ministerio de Salud, Instituto Nacional de Nutrición, Lima, Perú.

Coxe, A. and Coxe, Ar. (1987), *The Great Book of Sausages*, Buffalo, Paperback.

Delgado, J.V. Camacho, M.E., León, J.M., de la Haba, M.R., Vallecillo, A., Barba, C. and Cabello, A. (2004), 'Poblaciones porcinas de Iberoamérica', in: J.V. Delgado Bermejo (coord.ed), *Biodiversidad porcina iberoamericana. Caracterización y uso* sustentable, Servicio de Publicaciones de la Universidad de Córdoba. Córdoba, pp. 21–31.

Erbguth, F.J. (2004), 'Historical Notes on Botulism, *Clostridium botulinum*, Botulinum Toxin, and the Idea of the Therapeutic Use of the Toxin, *Movement Disorders*, vol. 19, Suppl. 8, S2–S6

FAO (2006), Retrieved March 2006 from http://faostat.fao.org/

Flores, J. (1997), Mediterranean *vs* northern European meat products. Processing technologies and main differences. *Food Chemistry*, vol. 59, pp. 505–510.

Giacosa, I.G. (1992), *A taste of ancient Rome*, Chicago, The University of Chicago Press.

González-Méndez, M.F., Zamorano-García, L., Cumplido-Barbeitia, L.G. y González-Ríos, H. (2002), Secado de carne para la elaboración de "Machaca", in: J.A. Pérez-Álvarez, J. Fernández-López and E. Salas-Barberá (eds), *Industrialización de productos de origen animal*. Ed. Escuela Politécnica Superior de Orihuela, Universidad Miguel Hernández. Elche, pp. 147–160.

Issanchou, S. (1996), 'Consumer expectations and perceptions of meat and meat product quality', Meat *Science*, vol. 43, pp. 5–19.

Leistner, L. and Rodel, W. (1976), The stability of intermediated moisture food with respect to microorganisms, in: R. Davies, G.G. Birch and J. Parker (eds), *Intermediate Moisture Foods*, London, Elsevier Applied Science Publishers, pp. 120–132.

LeMaguer, M. and Jelen, T. (1986), *Food Engineering and Process Applications*, London, Elsevier Applied Science Publishers.

Leroy, F., Verluyten, J. and Vuyst, L. de (2006), 'Functional meat starter cultures for improved sausage fermentation', *International Journal of Food Microbiology*, vol.106, pp. 270–285.

Mata, L.J., Campos, M., Basso, E., Compagnucci, R., Fearnside, P., Magrin, G., Marengo, J., Moreno, A.R., Suarez, A., Solman, S., Villamizar, A., Viller, L. (2001), 'Latin America', in: McCarthy et al. (eds), *Climate Change 2001: Impacts, Adaptation, and Vulnerability*, Cambridge, Cambridge University Press, pp. 693–734.

Monín, A. (1991), Chacinados caseros. 2da edición. Editorial Albatros. Buenos Aires.

Norman, G.A. and Corte, O.O. (1985), 'Dried salted meats: charque and carne-de-sol', *FAO Animal Production and Health Paper 51*. Food and Agriculture Organization of the United Nations. Rome, at http://www.fao.org/docrep/003/x6555e/X6555E00.htm #TOC.

Panati, C. (1989), *Patani's Extraordinary Origins of Everyday Things*, New York, Perennial Library.

Pardi, M.C. et al. (1996), *Ciência, Higiene e Tecnologia da Carne*. 1 ed. v. 2. UFG Goiania, pp. 588–1110.

Pearson, A.M. and Tauber, F.W. (1984), *Processed meats*. 2nd edition, Westport, Connecticut, Avi Publishing Company.

Peterson, M.S. (1963), 'Factors contributing to the development of today's food industry', in: M.S. Peterson and D.K. Tressler (eds), *Food technology the world over*, Westport, Connecticut, Avi Publishing Company. pp. 45–68.

Pinto, M.F., Pnsano, E.H.G., Franco, B.D.G.M. and Shimokomaki, M. (2002), 'Charqui meats as fermented meat products: role of bacteria for some sensorial properties development', *Meat Science*, vol. 61, pp. 187–191.

Ray, F.K. (1994), Oklahoma Cooperative Extension Fact Sheets, Retrieved March 2006, from http://pods.dasnr.okstate.edu/docushare/dsweb/Get/Document-2055/F-3994web.pdf.

Resurreccion, A.V.A. (2003), 'Sensory aspects of consumer choices for meat and meat products', *Meat Science*, vol. 66, pp. 11–20.

Reyes-Cano, R., Dorantes-Álvarez, L., Hernández-Sánchez, H. and Guitérrez-López, G.F. (1994), 'A traditional Intermediate Moisture Meat: Beef Cecina', *Meat Science*, vol. 36, pp. 365–370.

Shepard, S. (2000), *Pickled, Potted and Canned: How the Art and Science of Food Preserving Changed the World*, New York, Simon and Schuster, pp. 112–114.

Sobal, J. (1999), Social change and food ways. Symposium on Cultural and Historical Aspects of Foods – Yesterday, Today, and Tomorrow. Corvallis, Oregon.

Torres, E.A.F.S., Shimokomaki, M., Franco, B.D.G.M., Landgraf, M., Carvalho Jr, B.C. and Santos, J.C. (1994), 'Parameters determining the quality of Charqui, and intermediate moisture meat product', *Meat Science*, vol. 38, pp. 229–234.

Wright, C.A. (1999), *A Mediterranean Feast: The Story of the Birth of the Celebrated Cuisines of the Mediterranean from the Merchants of Venice to the Barbary Corsairs*, New York, Morrow Cookbooks.

Youssef, E.Y., García, C.E.R. and Shimokomaki, M. (2003), 'Effect of salt on color and warmed over flavor in charqui meat processing', *Brazilian Archives of Biological Technology*, vol. 46, pp. 595–600.

Zapata, M., Bañon, S. and Cabrera, P. (1992), *El pimiento para pimentón*, Madrid, Ediciones Mundi-Prensa.

Zeuthen, P. (1995), 'Historical aspects of meat fermentation', in: G. Campbel-Platt and P. E. Cook (eds), *Fermented Meats*, Suffolk, Blackie Academic and Professional, pp. 53–68.

Zumalacárregui, J.M. and Díez, V.A. (2001), 'Cecina y tasajo' in: S. Martín Bejarano (ed.) *Enciclopedia de la carne y de los productos cárnicos*, Plasencia, Martin and Macias, pp. 1163–1176.

Chapter 8
Market Integration and Market Power in the Internal EU Market for Butter

Heinrich Hockmann and Éva Vőneki

Introduction

The creation of the internal markets and the introduction of the euro were intended to foster the exchange of goods, services and inputs between the Member States (MS) of the European Union. These processes implied a liberalization of intra-EU trade and more intense competition among firms through market integration. This was assumed to foster specialization and to improve the efficiency of agricultural production within the EU.

Whether these contributions could be exploited depends strongly on the characteristics of the markets, i.e. the extent of market power, the existence of barriers to entry, product differentiation, and so on. In this chapter we concentrate on the EU dairy market, and investigate whether an internal market with intense market integration for this produce exists. Our considerations are restricted to the developments of the butter market. The reasons for these choices are that: (1) milk production is an important sector in EU agriculture; and (2) butter is a homogeneous good which suggests that trade patterns are not biased by product differentiation or other instruments to circumvent the negative effects of intense competition as can be assumed for cheese and fresh milk products.

In this chapter the following questions will be discussed:

1. What is the extent of market integration in the EU-15? In this context it will be analysed whether there is any indication of insufficient price transmission among the EU countries and, if yes, what are the causes of this phenomenon. In addition, it will be investigated whether structural breaks can be identified, and whether these can be traced back to major changes of the institutional environment, change such as the introduction of the euro or the enlargement of the EU in 2004.
2. How does market integration perform in peripheral regions? In this context the experiences of Hungary during the accession process will be investigated. This part focuses on the identification of changes in patterns of price transmission between Hungary and selected EU-15 Members.

These problems will be studied using time-series models, especially co-integration techniques. Within this framework market integration is indicated by the joint movements of the national product prices. One advantage of this approach is that compared with alternative econometric models the data requirements are relatively low, while at the same time relatively strong conclusions can be derived from the estimates.

The structure of the chapter is as follows: Section 8.2 deals with the trade pattern of butter on the international as well as the internal market. The Common Market Organization for milk and milk products is described in Section 8.3. Butter price developments in selected countries are the topic of Session 8.4. Section 8.5 outlines the econometric approach, while the results of the empirical analysis are presented in Section 8.6. Finally, Section 8.7 summarizes our arguments. Based on the descriptive and econometric analysis, the two questions posed above will be answered.

8.1 Developments on the EU market of dairy products

Trade of dairy products

With a share of about 30 per cent on the world market, the EU is the most important exporter of dairy products. Exports amounted to 15 ml t milk equivalents in 2004 which is 11 per cent of EU milk production. The destinations of European exports are Russia, Northern Africa, the Near East and the USA. Other large exporters are New Zealand, Australia and the USA with a market share of 21 per cent, 13 per cent and 7 per cent, respectively. Australia and New Zealand export a large share of their dairy produce. Exports from the EU, on the contrary, represent only a small share of its total output. Because of the production quota and almost stable domestic consumption, EU exports are also relatively stable. However, since 1995, the EU international market shares have been on the decline for almost all dairy products, while New Zealand has contributed increasingly to world exports. The EU is not only the most important exporter but also an importer of dairy products. It imports especially butter and skimmed milk powder (SMP) from New Zealand and cheese from the Switzerland. With 3 ml t milk equivalents, imports account for only a relatively small percentage of cheese and butter consumption in the EU (Salomon, 2003; Rabobank, 2004; FAOSTAT, 2005).

However, intra-EU trade of dairy products is much more pronounced than trade with third countries (Figure 8.1). In 2004, the share of exports to third countries in total production amounted to 16 per cent, 26 per cent and 9 per cent for butter, SMP and cheese, respectively. The corresponding figures for intra-EU trade were 27 per cent, 50 per cent and 38 per cent (EUROSTAT, 2006a). Intra-EU trade of dairy products was relative stable in the last decade. However, as a result of WTO restrictions on subsidized exports, the share of intra-EU exports shows an increasing trend which can be considered as an indicator of the ongoing specialization of dairy production within the EU.

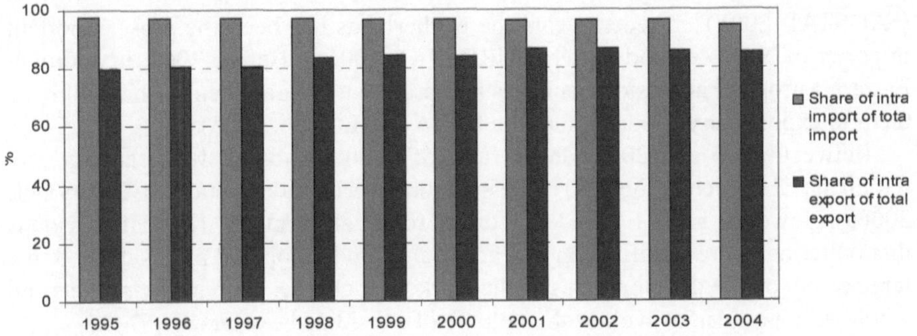

Figure 8.1 Share of intra-trade in total trade of dairy products, 1995–2004
Source: EUROSTAT (2006a).

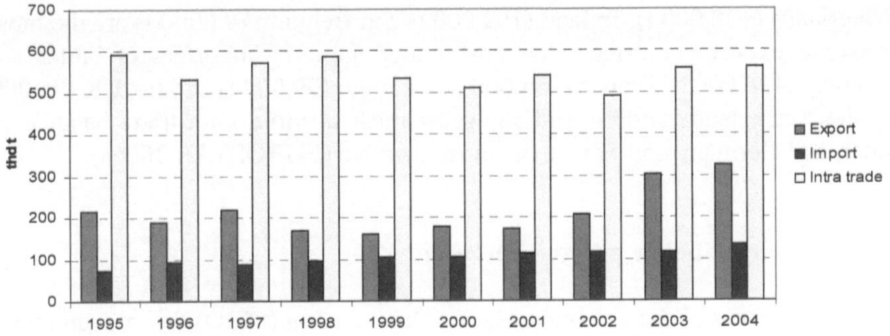

Figure 8.2 Intra- and extra-EU butter trade, 1995–2004
Source: EUROSTAT (2006a).

8.1.1 Butter trade

In milk equivalents, butter is the most important dairy product traded on the world market; the share of butter in total world trade of dairy products was about 50 per cent in the last decade and amounted to about 1ml t (FAOSTAT, 2005). The supply side is rather concentrated. The main exporters are the EU, New Zealand and Australia. In 2004 their market shares were 34 per cent, 38 per cent and 12 per cent, respectively (Rabobank, 2004). Since 1999, EU butter export has shown an increasing tendency, and Community exports doubled between 2000 and 2004 (Figure 8.2).

As in the case of dairy products in general, the import demand for butter is much more fragmented. For many years Russia was the largest importer (170,000 t per year on average), followed by the EU with about 130,000 t. Because of a preferential agreement that came into force when the UK joined the EU, a large proportion of these imports (87,000 t in 2004) consists of butter from New Zealand

(FAOSTAT, 2005). In recent years the Netherlands has been the most important importer of New Zealand butter (EUROSTAT, 2006a). Before 2004, in addition, Eastern European accession countries had preferential agreements which allowed them to export dairy products into the EU (Agrar Europe, 2005).

Between 1995 and 2004, on average, 550,000 t of butter was traded yearly within the EU which is about 50 per cent of total world trade in butter (EUROSTAT, 2006a). However, intra- EC trade fluctuated relatively strongly. The main reason is that butter has a 'rest-utilization character' and production and prices depend to a large extent on the developments of the market for cheese, fresh milk products and whole milk powder, as well as the volume of subsidized exports.

Almost all EU Member States export butter to third countries, while import plays an important role only in the Netherlands and the United Kingdom (Figure 8.3). The most important exporters are the Netherlands (65,000 t) and Belgium (28,000 t). The Netherlands and Ireland have the most positive, while the UK and France the most negative extra-trade balance of butter. Within the EU, the Netherlands (128,000 t), Ireland (102,000 t) and Belgium (92,000 t) are the most important export countries. The most important intra-importers of butter are Germany (136,000 t), France (135,000 t), Belgium (99,000 t) and the UK (96,000 t). The Netherlands and Ireland show the most positive intra-trade balance of butter, and Germany and France the most negative (EUROSTAT, 2006a).

8.2 EU market policy and the butter market

The core element of the Common Market Organization (CMO) for milk and milk products is the intervention system intended to remove structural surplus production of butter and SMP from the market (Agrar Europe, 2005). Intervention prices set by the EU for butter and SMP provide a price floor to the market and, as a result, have kept EU prices artificially higher than world market prices. However, with the

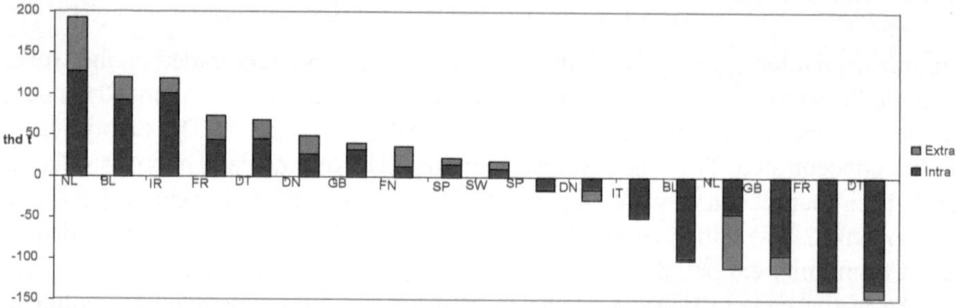

Figure 8.3 Intra- and extra-EU butter exports and imports by EU Member States (average for the period 2000–2004)

Source: EUROSTAT (2006a).

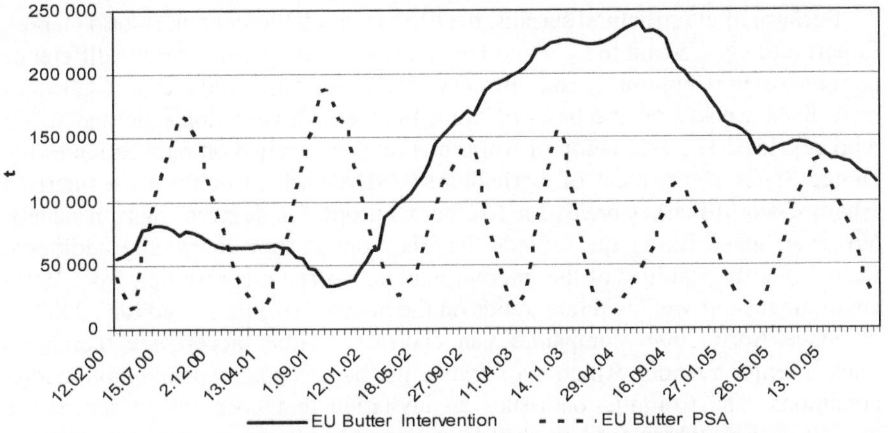

Figure 8.4 EU intervention stocks of butter

Source: ZMP (different issues).

CAP Reform in 2004, the EU is beginning to move away from production-based market support and to reduce the gap between EU and world market prices by cutting intervention prices and the guaranteed amount of intervention purchases.

After a period of stable prices, butter intervention prices were reduced by 7 per cent per year between 2004 and 2006, and by 4 per cent in 2007, giving a total reduction of 25 per cent by 2007. Intervention purchases can be made between 1 March and 31 August, if the reported market prices drop below 92 per cent of the intervention price for two consecutive weeks. The intervention agency can buy butter into intervention at 90 per cent of the intervention price. Until 2004, the amount of butter allowed into intervention was limited to 70,000 t. Between 2005 and 2008 the maximum amount will be reduced to 30,000 t. If the limits on quantity are reached, the Commission can decide to open a tender. Under this system, the Management Committee for Milk and Milk Products is asked to vote on the maximum buying-in price. Tenders above this accepted price are rejected. When market prices allow, intervention stocks can be sold, again by tender.

Another means of assistance for butter is private storage aid (PSA). PSA is intended to alleviate disruption to the market caused by seasonal fluctuations of butter production. PSA is usually available from 15 March to 15 August each year. Under PSA arrangements, butter remains in the ownership of the traders, but they receive a subsidy to cover at least part of the storage costs. Following the reform of the CAP, rates of aid paid are also being reduced.[1] The development of intervention stocks is shown in Figure 8.4.

1 The general rules for the application of intervention are laid down in Council Regulation 1255/1999. Detailed rules for the application of Butter Intervention and Private Storage Aid for Butter are laid down in Commission Regulation (EC) 2771/99.

Because of the structural surplus, the EU has to sell butter on the world market. Export refunds are paid to EU exporters in order to compensate for the difference between (high) Community and (lower) world prices. Until 2004, export subsidies were fixed weekly on the basis of the difference between domestic and world market prices. The estimation of world market prices relied on publication of the United States Department of Agriculture (USDA), which publishes regularly a range of world market prices for Northern Europe for different dairy products. However, when fixing the refunds the EU Commission considered additional factors like the stability of the internal market, restrictions resulting from WTO commitments, as well as future trends on the markets (Court of Auditors, 2003).

Since April 2004, companies can choose to either accept the 'common' rates, or enter a tender, which may or may not be accepted depending on market conditions. The Commission issues an invitation to tender, which specifies a closing date by which tenders must be received by the competent authorities in the Member States (MS). Exporters then bid for the level of refund and tonnage they require. The Commission adjudicates the tenders and proposes refund rates taking into account the same factors as for the common rates. The Commission proposes a maximum refund rate, which is voted on by the Milk and Milk Products Management Committee. A successful bid will be at or below the maximum rate set by the Committee. Successful tenderers receive the rate of export refund they individually bid for. This rate applies only to the quantity of product accepted in that bid. Export refunds set by tender are slightly higher than the refunds paid under the normal system.[2] Following the reform of the CAP and the need to fulfil WTO requirements, EU export refunds for butter were reduced from €185 in 2003 to €92 in December 2005 (Figure 8.5).

Complementary to the other instruments, border protection prevents unlimited imports to the EU. Import tariff levels are set by agreement within the World Trade Organization (WTO), and are currently being reviewed. The last WTO agreement set three types of tariffs: market access tariffs – the general level of tariff, open to any amount of product; minimum access/tariff rate quotas – a fixed amount of product (quota) that could be imported at a lower rate of tariff than under the market access tariff; current access tariffs are again fixed quotas of products at lower rates of tariff, but specifically attached to certain countries that have historical trading links with the EU. In the dairy sector the tariff rate quotas were 2 per cent of EU consumption, and the current access quotas around 3 per cent of EU consumption (Table 8.1).

2 The general rules for the application of export refunds are laid down in Council Regulation 1255/1999. Detailed rules for the application of the system of export licences and refunds for milk and milk products are laid down by Commission Regulation (EC) 174/1999. Details of the tender procedure for export refunds can be found in Commission Regulations (EC) 580/2004, 581/2004, and 582/2004. Details of the rules relating to Export Refunds that are differentiated by destination can be found in Commission Regulation (EC) 800/1999.

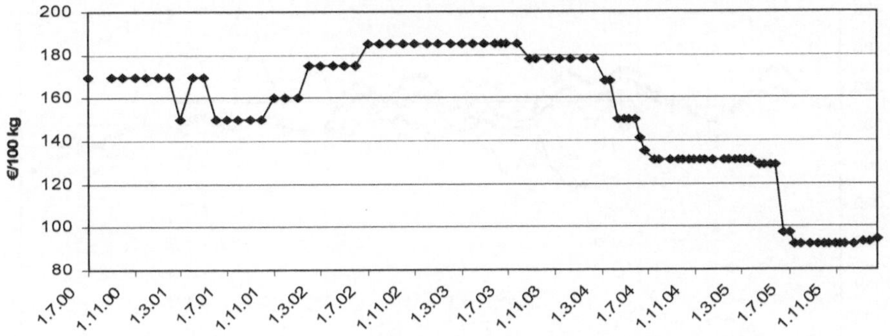

Figure 8.5 EU export refunds for butter

Source: MDC Datum (2005).

Table 8.1 Import tariffs of butter

	2000/01
Import tariff (€/t)	1,896
Minimum Access Volume (t)	10,000
Current Access Import Tariff Quota (t), New Zealand	76,667

Source: MDC Datum (2005).

8.3 Development of EU market prices of butter

The high protection of the EU milk market led to large differences between internal and external EU prices of dairy products (Figure 8.6). Moreover, EU intervention created an effective floor to the market and wholesale butter prices have oriented towards this price level. Correspondingly, butter was affected by the CAP reform in 2004, i.e. the reduction of the guaranteed amount of intervention, as well as the reduction of intervention prices. Additional pressure on the prices resulted from the decrease of export subsidies. There has also been a fall in subsidized sales to the food manufacturing market, and a release of interventions stock accompanied by a relatively high amount of production. Furthermore, lower prices in the new MS have contributed to the price reduction in the EU.

Besides these institutionally affected common movements of butter prices, structural differences between the countries can be observed. These reflect the trade patterns revealed in Figure 8.3. Germany as the largest importer has the highest butter prices. On the other hand, in Ireland, one of the largest intra-exporters, prices are significantly lower. In the Netherlands, which belongs to the group of large exporters as well as to the group of large importers, prices are at an intermediate level. Moreover, price differences seemed to disappear with the introduction of the CAP reform in 2004. Figure 8.6 additionally provides

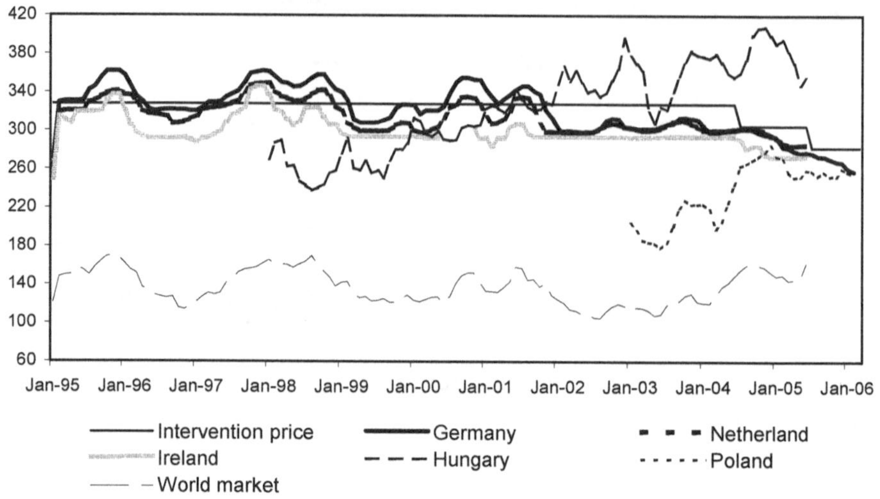

Figure 8.6 EU wholesale prices and world market prices of butter, for selected countries (€/100 kg), January 1995–March 2006

Source: EUROSTAT (2006b); USDA (2005).

information about butter prices in two New Member States (NMS). In Poland butter prices were significantly below the EU level before accession. However, prices increased and converged after accession and almost reached the level set by the intervention price. The development in Hungary was quite different. There butter prices have shown an upward trend since 1998. In 2000, they reached the EU level, and since 2002 butter prices have been significantly higher than in the EU. In addition, they appear to be unaffected by the change of the institutional framework resulting from the CAP reform, in 2004. This development is caused by the low level of butter production and consumption in Hungary. Moreover, production decreased in recent years because of the abolishment of butter subsidies. Since butter consumers' reactions are inelastic, prices could increase significantly.

8.4 Theoretical framework

The preceding descriptive analysis of price movements suggests that the developments of butter prices are strongly influenced by the regulatory environment set by the CMO for milk and milk products and by the market conditions prevalent in the individual countries. The empirical information can be summarized as follows:

1. Butter prices in the EU followed the development of world market prices, although on a significantly higher level.

2. Price differences among old MS corresponded to the trade patterns between the MS. Export countries have lower prices than import countries.
3. The market regulations are effective in the EU-15, i.e. the market prices follow the reduction of the institutional prices such as intervention prices and export refunds.
4. Prices in the NMS show heterogeneous developments not necessarily comparable with the development in the EU-15.

The preceding analysis allows no conclusion regarding the causal structure of butter prices, i.e. whether prices are affected by common factors only or whether there are strong relationships between the individual series. This problem of market integration is typically studied with time series models. One starting point is a vector autoregressive (VAR) model:

$$\mathbf{p}_t = \mathbf{a} + \sum_{i=1}^{\rho} \mathbf{A}_i \mathbf{p}_{t-i} + \mathbf{u}_t, \text{ with } \mathbf{p}_t = [p_{ger}\, p_{ned}\, p_{irl}\, p_{hun}], \quad (1)$$

where ρ represents the degree of autocorrelation of the time series and p_{ger}, p_{ned}, p_{irl}, and p_{hun} denote butter prices in Germany, the Netherlands, Ireland, and Hungary, respectively. The $\mathbf{A}_i$ are coefficient matrices to be estimated. **a** represents a function which may include constants, dummy variables, time trends, or other exogenous variables.

Since butter prices were not stationary, estimation of (1) would have led to spurious results regarding the significance of the parameters. In order to circumvent this problem (1) was transferred into a vector error-correction (VEC) formulation:

$$\Delta\mathbf{p}_t = \mathbf{a} + \Pi\mathbf{p}_{t-1} + \sum_{i=1}^{\rho-1} \mathbf{\Gamma}_i \Delta\mathbf{p}_{t-i} + \mathbf{u}_t, \text{ with } \mathbf{\Pi} = -(\mathrm{I} - \sum_{i=1}^{\rho} \mathbf{A}_i) \text{ and } \mathbf{\Gamma}_i = -\sum_{i=1+1}^{\rho} \mathbf{A}_i. \quad (2)$$

In this model $\mathbf{\Gamma}_i$ are (k×k) matrices representing short-run adjustments. The term $\Pi\mathbf{p}_{t-1}$ represents the long-run relationships. Given that the vector-autoregressive process governing (2) is non-stationary, the rank (r) of the matrix Π is smaller than k, the number of variables. In this case, Π can be written as the product of two (k×r) matrices α and β, each with rank r: $\Pi = \alpha\, \beta'$. β is called the 'co-integration matrix'. It transforms the non-stationary vector $\mathbf{p}_{t-1}$ into a stationary process. Thus, $\beta'\, \mathbf{p}_{t-1}$ represents the long-run relationship in the model. The loading matrix α reflects the velocity at which, after a shock, the system converges to the long-term equilibrium. Unfortunately, matrices α and β cannot be used to identify economic relationships. First, the rank of Π, and thus of α and β, has to be determined with statistical methods, and congruence between the structural model and the number of co-integration vectors in β is not guaranteed. Second, α and β are not unique. Every transformation with a nonsingular matrix $\mathbf{C}$ ($\Pi = \alpha\, \mathbf{C}\, (\beta\, \mathbf{C}'^{-1})'$) provides a new loading ($\alpha\, \mathbf{C}$) and a new co-integration matrix ($\beta\, \mathbf{C}'^{-1}$) (Lütkepohl, 2004).

Since the main focus will be on the identification of relationships between the variables, the estimates are used to reveal causal relationships and to construct the impulse response functions. These analyses rely on the transformation of (2) into a vector moving average (VMA) process (3), respectively (Breitung et al., 2004):

$$\mathbf{p}_t = \mathbf{p}_0 + \Xi \sum_{i=1}^{t} \mathbf{u}_t + \sum_{i=0}^{\infty} \Xi_i^* \mathbf{u}_{t-i}, \text{ with } \Xi = \boldsymbol{\beta}_\perp \left[\boldsymbol{\alpha}_\perp'(\mathrm{I} - \sum_{i=1}^{p-1} \boldsymbol{\Gamma}_i) \boldsymbol{\beta}_\perp \right] \boldsymbol{\alpha}_\perp', \tag{3}$$

with initial conditions ($\mathbf{p}^*_0$). $\boldsymbol{\alpha}_\perp$ and $\boldsymbol{\beta}_\perp$ are orthogonal complements of $\boldsymbol{\alpha}$ and $\boldsymbol{\beta}$. Ξ_i^* are transitory effects, and Ξ represents the long run impact of the shock in the residuals.

8.5 Empirical results

The empirical analysis considers butter price developments in Germany, the Netherlands, Ireland and Hungary. Other MS were not considered in order to save on degrees of freedom. However, regarding the old EU, the main export and import countries are considered. In order to identify patterns in price developments, relatively long time series are required. Data for Poland have only been available since 2003. Among the NMS, Hungary was an important trading partner of the EU-15 before accession. Compared with other NMS this may suggest relatively strong market integration already before accession, a fact that has to be considered in the interpretation of the results. The econometric analysis covers the years from 1998 to 2005.

Table 8.2 provides some statistical test results regarding the structure of the model. According to the Hannan-Quinn- and Schwarz-criterion, the optimal lag length of (2) is 2. The correct identification of unit roots is pivotal for the formulation of the error correction model. Since none of the various tests for unit roots is superior under all circumstances, different test statistics were analysed. The Augmented Dickey Fuller (Fuller, 1996) and the Breitung (2002) test provide unambiguous results, insofar as the hypothesis of unit roots with drift cannot be rejected. In principle, the conclusions are confirmed by the Philips and Perron (1988) test. Only for Germany was the hypothesis rejected. The λ_{trace} and λ_{max} test were used to determine the number of characteristics roots of Π (Enders, 2004). Both criteria provide that the series of butter prices are characterized by one co-integration vector.

In addition, it was tested whether a long-run relationship without structural breaks existed. Based on a concentrated likelihood function recursive eigenvalues, a τ – t-statistics was computed (Hansen and Johansen 1999). The test provides the existence of a stable long-run relationship. However, since inspection of the series suggests a reduction of the price level in the MS in 2002, a dummy variable was considered in **a**. In addition, it was tested whether a time trend increases the explanatory power of the model. Because no significant influences could be detected, no trend was considered. Moreover, the test results suggested that possible trends are already absorbed by differencing of the variables.

Table 8.2 Properties of the butter price time series

		P_{ger}	P_{ned}	P_{irl}	P_{hun}
existence of unit root with drift (5% level of significance)	ADF	H_0 not rejected	H_0 not rejected	H_0 not rejected	H_0 not rejected
	Phillips-Perron	H_0 rejected	H_0 not rejected	H_0 not rejected	H_0 not rejected
	Breitung	accept H_0	accept H_0	accept H_0	accept H_0
lag length	Hannan-Quinn	2			
	Schwarz	2			
# of co-integrating vectors	λ_{trace} λ_{max}	1			
stability (5% level of significance)	τ – t-statistics	no break in eigenvalue			

Source: Own estimates.

Table 8.3 Causality tests

		P_{ger}	P_{ned}	P_{irl}	P_{hun}
Instantaneous causality	χ^2	1.21	8.8*	12.89***	10.11**
Granger causality	F-Test	1.67	10.33***	6.88***	1.63

Source: Own estimates.

In the following, the causality structure revealed by the estimates is analysed. Two concepts are used: Granger causality (Granger, 1969) and instantaneous causality (Burda, 2001). Granger causality is present when knowledge about the previous realization of a series improves the predictability of other series. It was checked via restrictions in the ECM (2) that lead to an exclusion of the variables of interest in the VAR. Instantaneous causality refers to a situation, in which the predictability of a series is improved when the realizations of other series in the period to be predicted are already known.[3] Instantaneous causality was analysed by checking correlations amongst the residuals. It comprises direct and indirect effects, which means that for $k > 2$ no direct relationships can be identified (Breitung et al., 2004).

Table 8.3 shows that most of the time series provide a significant contribution regarding the predictability of the development of the total system. This holds for instantaneous causality, and to a lesser degree for Granger causality. Only for Germany were significant contributions regarding predictability for both causality

3 In order to present only direct effects, the analysis of instantaneous causality is restricted to the impact of one time series on all others.

concepts estimated, suggesting that German prices affect prices in the export countries only marginally. The same can be concluded for Hungary.

Parameter estimates of (2) are presented in Table 8.3. Noticeable is the structural difference in the significance of the parameter which concerns the short- and the long- run adjustment. The long-run effects are generally highly significant. On the contrary, high t-values for the short-run effects are rather an exception. This provides an indication that that second-order autocorrelation explains price variation only to a limited extent, and most of the adjustments will happen rather quickly. The dummy variable inserted to account for a possible structural break is significant in all equations. In Germany and the Netherlands prices drop, while in Ireland the dummy has a positive impact. Thus convergence of EU prices occurs. Moreover, it captured the period after the introduction of the euro, which in turn suggests that the euro has had a positive impact on the exchange of goods among MS, which may result from the decrease of transaction costs.

In addition, the R^2 statistics are reported in Table 8.4. A satisfactory adjustment could be achieved for Germany and the Netherlands. For Hungary and Ireland, the R^2 values are rather low suggesting that important influences like world market prices and development in other export regions have not been considered consistently.

According to the Jarque-Bera-test,[4] normality of the residuals cannot be rejected for Germany and the Netherlands at reasonable levels of significance. For Ireland skewness, while for Hungary kurtosis, of the residuals leads to a rejection of the normality hypothesis. Homoskedasticity of the residuals was checked with an autoregressive conditional heteroskedasticity maximum likelihood (ARCH-LM) test. The LM-values provide that homoskedasticity cannot be rejected except for Ireland. A further problem concerns autocorrelation among the residuals. The Portmanteau and the Breusch Godfrey test suggest that autocorrelation among adjacent residuals cannot be rejected. Graphical inspection provides that this mainly concerned the results obtained for Dutch butter prices.

Even if not all test statistics are fully satisfactory, model (2) can be regarded as a reasonable approximation of the developments in the EU butter market. Thus, further analysis of the estimates in the form of the impulse response function can be justified. According to the analysis of the residual, it cannot be excluded that the errors are (instantaneously) correlated. In order to isolate the direct effects, orthogonalized impulse responses were calculated with a Cholesky transformation of the variance-covariance matrix of $\mathbf{u}$ in (3). The matrix Ξ was adjusted correspondingly. This procedure has the disadvantage that impulse responses are affected by the ordering of the variables (Breitung et al., 2004). Various orderings have been compared. The typical patterns – negative or positive impact on one variable, as well as the size of the responses – remains stable except when the shock appears in the Netherlands first. In that case, innovation responses in the old MS reached the level observed for shocks in Ireland.

---

4 A description of the test statistics can be found in Lütkepohl (2004).

Table 8.4 Parameter estimates and residual analysis

		Δp_{ger}	Δp_{ned}	Δp_{irl}	Δp_{hun}
parameter estimates					
	$\Delta p_{ger}(t-1)$	0.134	-0.15*	-0.032	0.25
lagged	$\Delta p_{ned}(t-1)$	0.51***	0.76	0.23	0.61
endogeneous	$\Delta p_{irl}(t-1)$	0.33***	0.32***	0.12	-0.57
	$\Delta p_{hun}(t-1)$	-0.053*	-0.003	-0.02	0.12
constant		-8.67***	8.65***	7.71**	30.21***
dummy		-4.07***	-3.55***	3.21*	13.10**
loading matrix		-0.03***	0.03***	0.03**	0.10***
co-integration vector		7.51***	-9.45***	1.	-0.38**
R^2		0.65	0.66	0.14	0.13
residual analysis					
nonnormality (Jarque-Bera-test)		JB = 2.72; p =0.25	JB = 2.31; p =0.31	JB = 20,16; p = 0.00	JB = 10.07; p = 0.00
homoskedasticity (ARCH LM test)		LM = 18.8; p =0.27	LM = 10.4; p =0.84	LM = 29.5; p =0.02	LM = 10.8 p =0.81
autocorrelation		Portmanteau test		Q_{16} = 267.4; p =0.07	
		Breusch Godfrey test		LM_5 = 115.8, p=0.00	

Source: Own estimates.

Typical innovation responses are given in Figure 8.7. Some shocks in the residuals have a permanent impact on the other variables. One reason is the existence of unit roots, which, by definition, have a permanent impact on the change of the corresponding time series. In addition, permanent influences are also caused by the co-integration among the variables. The long-run effects of shocks are governed by Ξ in (4). Since the matrix has rank k-r, the number of zero columns cannot exceed r (Breitung et al., 2004). This implies that there are at most r shocks with transitory effects. This was only observed for Hungary as a reaction to shocks in Irish prices: Otherwise permanent impacts, albeit sometimes on a relatively low level are present. In the following discussion, only the impact of shocks in one country on the prices in other MS will be discussed.

The highest innovation responses could be observed for shocks in the Irish prices in the old MS. All effects are positive, suggesting that a reduction of competitiveness of Irish exports leads to higher prices in the EU. Shocks in Germany affect the butter prices in all countries, positively. Since Germany is the largest net-importer of butter from other EU countries, additional demand for butter leads to higher exports to Germany, which in turn leads to a greater scarcity of butter in these countries. Correspondingly, higher prices in the Netherlands lead to higher butter prices in Germany. However, Irish prices are affected negatively. Since both the Netherlands and Ireland are large butter exporters, the reaction may result from the competitive forces on the internal market. In that case, higher Dutch butter prices are associated with higher exports. This, ceteris paribus,

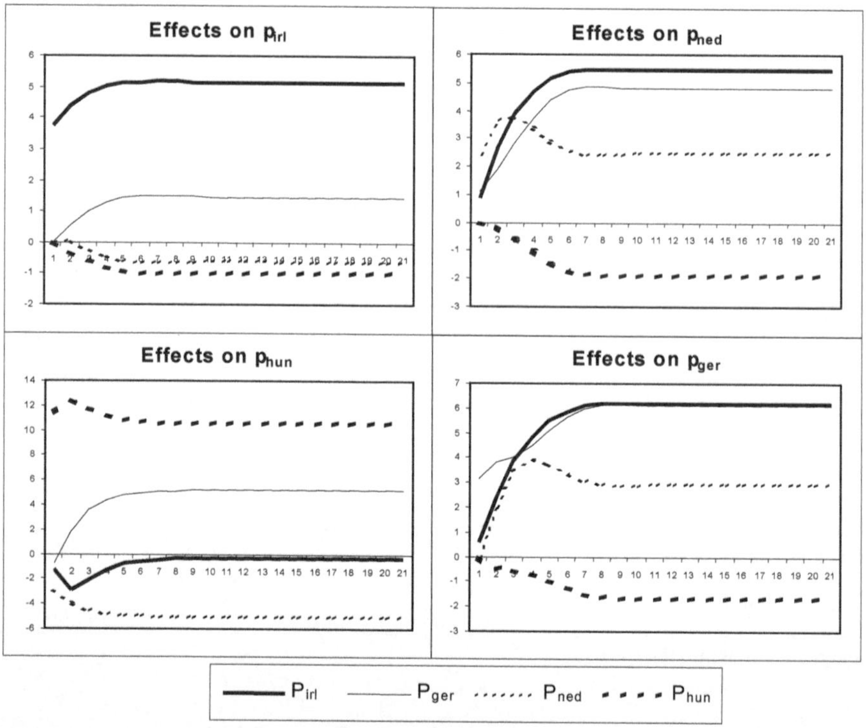

Figure 8.7 Orthogonal impulse response functions

Source: Own estimates.

increases the competitive pressure. Since Irish exports will decrease, more butter has to be sold on its domestic market. Because of its small population, the domestic market can absorb the additional quantities only when prices decrease significantly. However, the reaction of Dutch butter prices show that this effect is not symmetric. This asymmetry suggests that importers are differently positioned for exploring the markets in other EU countries. Changes in Hungarian prices have a negative impact on the market in the old MS. However, the effects are relatively low. Moreover, shocks on the Hungarian markets appear to be decoupled from the effects in the old MS. This suggests that this is a low degree of integration in the internal market, and that butter prices are still mainly determined by the development on the domestic market.

8.6 Conclusion

Inspection of price developments suggests that the EU butter market is highly integrated. This hypothesis is confirmed by the results of the econometric analysis.

Statistical tests indicate that the sample is characterized by one co-integration vector that captures long-run relationships amongst butter price developments in the different EU countries. Furthermore, the price differentials amongst the countries are consistent with the trade pattern of butter. Moreover, in the period under investigation no break in the co-integrating relationship could be revealed. This suggests that recent changes like the introduction of the euro in 2002 and the EU enlargement, as well as the CAP reform in 2004 have changed the degree of market integration significantly. However, the introduction of the euro may have affected the short-run relationships amongst the prices insofar as a convergence of European prices could be observed. One reason for this development in the decrease of transaction costs is the EU-internal exchange which has facilitated intra-EU trade of butter. The econometric analysis also shows that the butter trade between EU countries may not be affected by market power. On the one hand, export by countries is much more concentrated than imports, but, on the other, Germany as one of the largest producers of butter is also the largest importer. Strong relationships amongst butter prices were revealed by the causality tests and the impulse response functions. These analyses provide some indication that mutual price effects are more pronounced in adjacent countries. This is consistent with the finding of what are called the gravity models of international trade.

In addition, we analysed whether the accession to the EU has changed price formation in the NMS. The results are not unambiguous. Before accession, butter prices in Poland were lower than in the EU. However, during the pre-accession period a catching-up process occurred, and now butter prices are almost at the same level of the old MS. The experience of Hungary is different. In that country, initially prices were also lower than EU prices. However, prices have increased there since 1998, and already before accession they had reached a higher level than within the EU. Moreover, the impulse response analysis shows that shocks have different effects in Hungary than in other EU countries. Only with regard to Germany could consistent price changes be observed. This result corresponds to the finding that especially between adjacent countries relatively strong ties exist amongst prices. Altogether this suggests that the Hungarian butter market is less integrated into the EU market than into the markets in the old MS. However, the econometric analysis suggests that Hungarian prices are part of the co-integration relationship. This contradiction is solved by taking the relationships to the German market into account.

Although the model discussed in this chapter provides some important insights, the statistical properties were not totally satisfactory. This calls for some refinements in the estimation. These include the consideration of price series for other countries, as well as of further exogenous variables, in order to increase the statistical adjustment. Moreover, having more countries would also allow more co-integrating relationships to be detected. This would be an indicator that there are not only one but more long-run relationships, and that market integration may work differently in different regions.

References

Agrar Europe (2005), CAP Monitor, London.

Breitung, J. (2002), Nonparametric Tests for Unit Roots and Cointegration, *Journal of Econometrics*, vol. 108, pp. 343–364.

Breitung, J., Brüggemann, R. and Lütkepohl, H. (2004), 'Structural Vector Autoregressive Modeling and Impulse Responses', in: H. Lütkephohl and M. Krätzig, (eds) *Applied Time Series, Econometrics*, Cambridge, pp.159–196.

Burda, M. (2001), 'Testing for Causality with Wald Tests under Nonregular Conditions', Dissertation, Wirtschaftswissenschaftliche Fakultät. Humboldt – Universität zu Berlin.

Court of Auditors (2003), Special Report No 9/2003 concerning the system for setting the rates of subsidy on export of agricultural products (export refunds), together with the Commission's replies (2003/C 211/01), at http://www.eca.eu.int/audit_reports/special_reports/docs/2003/rs09_03en.pdf

Enders, W. (2004), *Applied Time Series Analysis*. 2nd edition, New York.

EUROSTAT (2006a), External trade, at http://epp.eurostat.cec.eu.int

EUROSTAT (2006b), Agricultural Markets – Prices, at http://epp.eurostat.cec.eu.int

FAOSTAT (2005), Trade Indices, at http://faostat.fao.org/faostat

Fuller, W.A. (1996), *Introduction to Statistical Time Series*. 2nd edition, New York.

Granger, C. W. (1969), 'Investigating Causal Relationships by econometric models and cross spectral models', *Econometrica,* vol. 16, pp. 121–130.

Hansen, H. and Johansen, S. (1999), 'Some tests for parameter constancy in cointegrated VAR models', *Econometrics Journal*, vol. 2, pp. 306–333.

Lütkepohl, H. (2004), 'Vector Autoregressive and Error Correction Models' in: H. Lütkephohl and M. Krätzig (eds), *Applied Time Series Econometrics*, Cambridge, pp. 86–158.

MDC Datum (2005), 'Milk Prices, Processor Data and Prices', at http://www.mdcdatum.org.uk

Phillips, P.C.B. and Perron, P. (1988), 'Testing for a Unit Root in Time Series Regression', *Biometrica*, vol. 75, pp. 335–346.

Rabobank (2004), Global Dairy Trade – a changing landscape, at http://far.rabobankamerica.com/app/content/fa_document/item/173

Salamon, P. (2003), *Weltmarkt für Milch und Milchprodukte im Spannungsfeld von EU Agrarmarktpolitik und WTO-Verhandlungen*, Institut für Marktanalyse und Agrarhandelspolitik an der FAL Braunschweig.

USDA (2005), 'Dairy Market *News*', *http://www.fas.usda.gov/dlp/circular/2005/05-07Dairy/toc.htm*

ZMP (different issues), *ZMP Marktbilanz.*

Chapter 9

The Romanian Food Consumption Model in the Context of European Union Integration

Cecilia Alexandri and Cornelia Alboiu

Introduction

Until recently, the food demand theory has been considered a particular case of general demand theory. Neo-classical economic theory was applied to all consumption goods on a non-differentiated basis, including food commodities. This approach that highlights the role of incomes and prices in explaining food behaviour continues to be largely used by economists (Senauer, 1997), even though the consumer's food behaviour in Western society has experienced a fundamental change. Furthermore, it is applied to the developing countries in order to explain the evolution of food consumption, although the economic and social context is fundamentally different here. Neo-classical theory considers foodstuff as a 'commodity', although in the developing and less-developed countries self-consumption represents an important part of food consumption (sometimes more than 50 per cent), (Colman and Trevor, 2000). In fact, a dichotomy can be found between food behaviour in the developed and less-developed countries.

In the Western food consumption pattern, productivity increases along the agri-food chains and induces food abundance, determining homogeneous food behaviour and convergences that are specific for mass consumption societies and a medium satisfaction of energy needs (Alexandros, 1995). In these countries, in effect, a quantitative saturation of consumption has been produced, while the budgetary restriction is not important in the case where food expenditures do not exceed 10–20 per cent of the family budget. For these reasons, the rich consumer acts less under the impulse of economic rationality, with nutritional rationality most often being in the first place. In this situation, the neo-classical theory of consumer behaviour optimization becomes less applicable in food consumption demand forecasting, at least in the rich countries. Many economists (Padilla, 1996; Schmitz, 1994) have investigated the problem whether food consumption in the developed countries is determined by real income, one of the conclusions being that 40–60 per cent of the food consumption is not dependent upon income. In the developing countries and some of the transition countries, the food consumption patterns have a heterogeneous structure that encompasses all the stages of the historical agri-food development. In many of them, the pre-agricultural food economy co-exists with subsistence agriculture based upon self-consumption, with forms of commercial agriculture that emerged with urbanization, and, finally, with an agro-industrial economy involved in international transfers. The population of these countries is divided into differentiated consumer groups, subject to specific restrictions that result in various kinds of food consumption behaviour and typologies (Hertel et al., 1997).The agri-food systems in these countries are neither unitary nor homogeneous, and they cannot be compared with any period in the history of Western economic development.

Romania, as a result of the high share of its rural population (45 per cent of the total population), is characterized by a mixed food consumption pattern. Thus, there the urban population has a consumption pattern in which access to food is mainly restricted by household purchasing power, and a consumption pattern of the rural population that includes the families that own land, whose food situation depends both upon their own production and their purchasing power. These consumption patterns do not exist in a pure state, as even the urban population has a significant degree of self-consumption, coming from the transfers of products from their relatives who live in the rural area. In this context, the objective of the chapter is to present the Romanian food consumption pattern by socio-economic characteristics (urban versus rural). The chapter also investigates the effects of income on consumption, taking into consideration Engel's Law.

The chapter is organized as follows: in Sections 9.1 and 9.2 the evolution of the main factors that influence the food consumption in Romania (income and food prices) is investigated. In Section 9.3 an overview of the structure of the households' expenditures in Romania is presented, followed by a presentation of the main characteristics of the Romanian food consumption pattern, differentiated by urban/rural areas (Section 9.4). The chapter ends in Section 9.5 with some concluding remarks.

9.1 Income distribution

Throughout the investigated period, the consumer price index significantly exceeded the nominal wages index; this policy was justified by the need to calm down inflation and create the conditions for macroeconomic stabilization. The result was certainly the diminution of real incomes: for example, average wages in the year 2004 were around 80 per cent of the 1990 level (Table 9.1).

In order to assess all forms of income at the household level, data from the Integrated Household Survey (IHS) initiated by the National Commission for Statistics (NCS) were used. The main indicator used by IHS for the assessment of household income is Total Nominal Income. The types of income included in Total Nominal Income are: wages, agricultural income, income coming from self-employment, social security income (pensions, unemployment wages, child allowances, etc.), income in kind and other income.

Table 9.1 Dynamics of average wages and pensions (1990=100%)

	1991	1993	1995	1997	1999	2001	2002	2003	2004
Average wages	81.7	59.4	66.5	56.3	56.0	62.4	63.9	70.8	78.3
Average pension of state social insurance pensioners	77.5	56.5	61.2	49.7	47.6	46.6	48.2	51.6	47.7

Source: Romanian Statistical Yearbooks, 1993–2005, National Institute of Statistics (NIS).

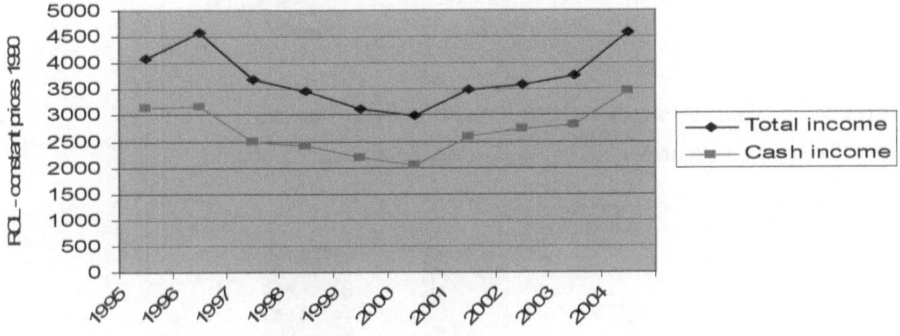

Figure 9.1 Total income and cash income per household

Source: Based on data provided by the National Institute of Statistics (NIS).

At the same time, the Total Nominal Income of each household is divided into two components: cash income, and income in kind (in the form of products). The total income of households began to increase from 2001, coinciding with the beginning of the period of economic growth; in 2004 these increases were 12 per cent higher compared with their 1995 level (Figure 9.1). Cash income represents 70–75 per cent of household total incomes.

The increase of income per capita is stronger in the urban areas compared with the rural areas (Figure 9.2). The level of incomes by rural/urban area reveals higher incomes in the urban area, as well as an increase in the income gap between the urban and the rural area in the last seven years (the per capita income in towns was 5 per cent higher than in the rural area in 1997, and 24 per cent higher in 2004). This finding, together with the fact that the food consumption level in the rural area is higher than in the urban area (2613 kcal in the rural area compared with 2390 kcal in the urban area, in 2004), reveals that the food consumption level depends to a lesser extent upon cash income in a subsistence economy.

In the income structure of rural/urban areas, cash income represents 88 per cent and incomes in-kind 12 per cent in the urban areas, while in the rural areas cash income accounts for 58 per cent and income in kind 42 per cent (Figure 9.3). The structure of incomes by sources of origin indicates that the income in-kind represents the most important income source for the peasant households. The

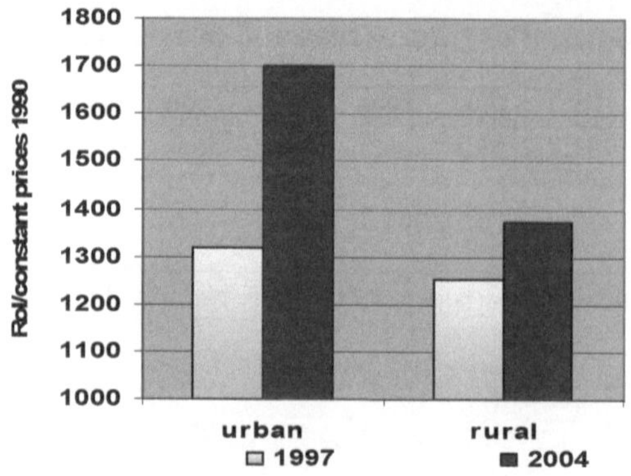

Figure 9.2 Incomes per capita by urban/rural areas

Source: Based on data provided by the National Institute of Statistics.

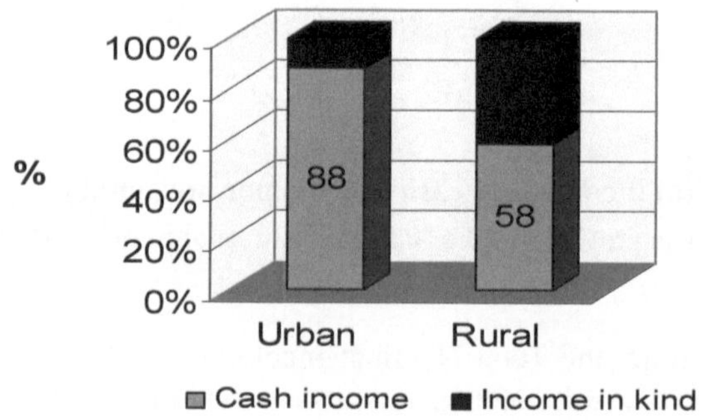

Figure 9.3 Structure of incomes by urban/rural areas

Source: Based on data provided by the National Institute of Statistics.

classification by groups of income (deciles) according to the head of household's occupational status reveals a concentration of employers and employees' households in the higher deciles, peasants and unemployed households in the lower deciles, and the households of pensioners in the middle deciles 4–7.

9.2 Food prices

In the year 2004, compared with 1990, general prices increased 2,375 times, while nominal wages by only 1,769 times (Table 9.2).

Since 1999, the food price index has been lower than the consumer price index (CPI). It should be mentioned that 1997 was the year when food prices were completely liberalized. The evolution of the real price index indicates the

Table 9.2 Consumer price index and nominal wages index (1990=100%)

	1991	1993	1995	1997	1999	2001	2002	2003	2004
Consumer price index (CPI)	270	2,987	9,353	33,077	76,728	150,290	184,162	212,291	237,504
Food products	286	3,361	10,469	35,886	68,098	132,797	157,096	180,131	197,195
Non-food products	268	2,907	8,776	30,823	75,233	144,158	180,930	210052	237774
Services	236	2,249	8,051	32,709	115,609	240,865	305,326	350645	402279
Nominal wages index	221	1,765	6,246	18,679	45,002	89,242	111,994	143,041	176,934

Source: Romania's Statistical Yearbook, 1997 and 2005.

diminution of real prices for food products and the spectacular increase of the real price index for services. The expenditures for house utilities (water, heating, electric power, etc.) have a high share in total services. The monthly expenditures for dwelling maintenance (house utilities) have increased so much that in many situations they exceed the minimum wages per household, the urban population being mostly affected by this situation.

When investigating the evolution of food prices, we find that the largest price increase in the post-liberalization period was experienced by cereal products (more than 1,500 times), meat products (more than 1,000 times), dairy products (more than 1,270 times); the group with the lowest price increase was represented by alcoholic drinks (over 730 times), in the year 2000 compared with 1990.

Because of the stronger price increase in basic products, which have the largest share in the household budgets (bread, meat, dairy products), consumption was affected as regards its quantity and particularly its quality, being deficient mainly in products of animal origin. The increase of prices for milling and baking products affects the population's living standard as these products have the largest share in the population's food consumption (the cereal-based products provide almost half of the food consumption expressed in calories).

According to a number of recent studies (Csaki and Kray, 2005), the Common Agricultural Policy will lead to an increase in domestic agricultural prices and thus to an increase of the food prices in the medium term. The most important increase of the medium- and long-term prices will be for eggs (by 16 per cent), and milk and dairy products (by 6.8 per cent). Although these price increases are expected to diminish the Romanian consumers' purchasing power, one could estimate that the impact will be modest as a result of the growth of the competition on the retail market. The increase of food prices will more severely affect the poorest households in the urban area, where the share of food consumption expenditures reaches 70 per cent of the consumption expenditures, while self-consumption is almost negligible.

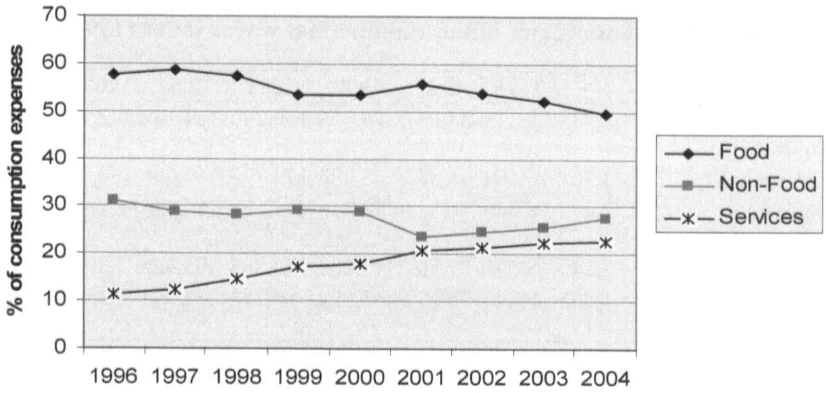

Figure 9.4 Structure of consumption expenditures

Source: Romania's Statistical Yearbook, 1997 and 2005.

9.3 Household expenditures

In total households, total expenditures are very close to the level of total incomes, so that savings account for only 1 per cent of total incomes of the average per household. Cash expenditures prevail in total expenditures, except for the peasant households. Thus, in 2004, cash expenditures represented 75 per cent on the average, while self-consumption represents 25 per cent of total expenditures. By residence area, the share of cash expenditures in total expenditures was 88 per cent in the urban area and 58 per cent in the rural area. This indicates a much lower demand for most products in the rural areas compared with the urban areas.

The structure of consumption expenditures reveals quite a high share of food consumption, ranging from 58 per cent in 1996 to 50 per cent in 2004 (Figure 9.4).

By residence area, food expenditures represent 44.5 per cent of total consumption expenditures in the urban area and 58 per cent in the rural area (year 2004). At the same time, the structure of consumption expenditures by income groups indicates a 72 per cent share of food expenditures for decile 1, and 34 per cent for decile 10, out of the total consumption expenditures in the year 2004.

As a consequence, in Romania not even the households with the highest incomes have a structure of consumption expenditures comparable to the Western-European average, where food expenditures do not exceed 20 per cent of total expenditures. The share of foodstuffs coming from the household's own production is minimum in employer households (29 per cent) and maximum in peasant households (70 per cent of food consumption expenditures). At the same time, it is worth mentioning the extremely low share of food consumed in restaurants (less than 1.5 per cent).

9.4 Food consumption

The Romanian food consumption pattern has certain characteristics that can be found in the behaviour of both the rural population and the urban population, though of course in different percentages, for instance:

- High consumption of bread and cereal products, representing the basic foodstuffs for all household types. Cereals provide 44 per cent of the calories in population's diet (47 per cent in the rural area), which implies a low diversification of food;
- Increase in the consumption of potatoes, fruits and vegetables in recent years;
- Increase in meat consumption in recent years, but it is still low, compared with the European average. Pork and chicken meat account for 75–80 per cent of total fresh meat consumption. However, the differences are quite significant among different household types: the households of employees, employers and unemployed eat mostly pork, while the families of pensioners and peasants mostly chicken;
- Increase of milk and dairy products consumption, mainly due to the increase in domestic production;
- Consumption increase of soft drinks, beer and wine;
- Very high share of self-consumption for certain products: namely, eggs, milk, wine and brandy (Figure 9.5).

The share of self-consumption in the rural area is extremely high for fresh meat (74.0 per cent), milk and cheese (about 70.0 per cent), eggs (90.0 per cent), fruit

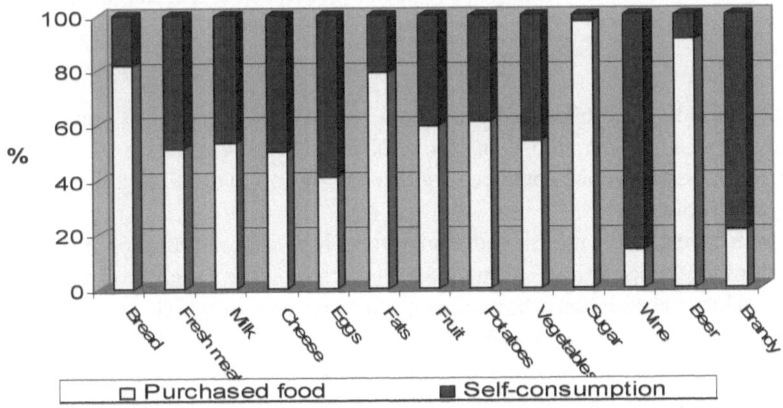

Figure 9.5 Origin sources of food consumption, 2004

Source: Based on data provided by the National Institute of Statistics (NIS).

Table 9.3 Food consumption by areas of residence in 2004 (annual average/capita)

Product	Unit of Measurement	Total	Urban	Rural	Urban-Rural
Bread	Kg	118.5	117.0	120.2	-3.2
Cornflour	Kg	15.6	7.8	24.6	-16.8
Fresh meat, of which:	Kg	30.5	32.8	27.9	4.9
Beef	Kg	4.0	5.0	2.8	2.2
Pork	Kg	10.3	10.8	9.7	1.0
Poultry meat	Kg	13.2	14.0	12.2	1.9
Meat-based products	Kg	11.3	11.5	11.1	0.4
Milk	L	71.2	62.5	81.1	-18.6
Cheese	Kg	13.5	13.7	13.3	0.4
Eggs	Pcs	161.1	156.3	166.7	-10.4
Fats	Kg	15.7	15.7	15.8	0
Fruit	Kg	29.5	34.1	24.2	9.9
Beans	Kg	5.5	4.4	6.7	-2.3
Potatoes	Kg	48.3	47.2	49.6	-2.3
Vegetables	Kg	85.2	86.5	83.8	2.6
Sugar (without sugar-based products)	Kg	10.0	10.2	9.8	0.4
Soft drinks	L	32.8	46.1	17.6	28.5
Alcoholic drinks, of which:	L	25.2	20.3	30.9	-10.6
Wine	L	13.2	8.4	18.6	-10.3
Beer	L	8.4	9.6	7.1	2.5
Spirits	L	3.1	1.7	4.6	-2.9

Source: Coordinates of the living standard in Romania. Incomes and consumption of the population, NIS (2004).

and vegetables (over 60.0 per cent), and alcoholic drinks (74.0 per cent). The lowest share of self-consumption can be observed in sugar, fats, soft drinks, beer and bread. The differences between food consumption in urban and rural areas can be seen in Table 9.3.

As we have already mentioned, owing to its high share of rural population (45.0 per cent of total population), Romania is characterized by a dual food consumption pattern. Thus, there is an urban consumption pattern, in which access to food is mainly restricted by the household's purchasing power, and the rural pattern, which includes the land-owning families whose food consumption depends both on self-consumption and on their purchasing power, determined mainly by their cash income. By comparing food consumption in the rural areas and in the urban areas, one can observe higher net values for the following products in the case

Table 9.4 Share of the calorific intake for the main food products consumed, by area of residence, 2004

	Urban (%)	Rural (%)	Total households (%)
Total calories, out of which:	100	100	100
Cereals total	42.2	46.5	44.3
bread	19.9	28.1	29.0
cornflour	3.1	9.0	6.0
Fresh meat and meat-based products	9.9	8.1	9.0
Fish	0.5	0.4	0.5
Milk total	4.4	5.6	5.0
Cheese and sour cream	4.1	3.5	3.8
Fats, of which:	16.2	15.0	15.6
edible oil	12.3	9.7	11.0
Eggs	1.5	1.5	1.5
Sugar and sugar-based products	7.6	6.0	6.9
Potatoes	3.9	3.7	3.8
Vegetables	3.0	2.7	2.9
Fruit	2.5	2.7	2.1

Source: Coordinates of the standard of living standard in Romania. Incomes and consumption of the population, NIS, 2004.

of the rural population's consumption: cornflour, milk, eggs, white beans and alcoholic beverages.

The higher consumption of cornflour and milk originates in the traditional food consumption pattern of the Romanian population, based on milk and cereal porridges (originally made of millet and later on maize). The rural households spend less money on food, yet in terms of quantity, their daily food consumption expressed in calories is higher than the consumption in the urban area. At the same time, there are quality differences between the food consumption in the two areas. Table 9.4 reveals the prevalence of cereals in the Romanian food consumption pattern of residence, in particular for the rural population, where about half of the daily number of calories originates from cereals.

Thus, although the rural population seems to have a more favourable situation as regards food quantity, the quality of their food intake is more deficient, because it is less diversified, with a high proportion of animal fats and eggs (rich in cholesterol), and a lower consumption of fruit. It is also worth noticing the consumption of alcoholic beverages, which is almost double that in the urban area; this is generally associated with a lower employment level and with poverty. These beverages are mainly produced on the peasant households and some of them are traditional Romanian products, i.e. certain specific wine varieties ('Feteasca Neagra', 'Busuioaca de Bohotin') and certain alcoholic beverages obtained from plums or other fruits (tzuica, palinca, etc.). Many other products are also produced

Table 9.5 The population's food consumption pattern on a comparative basis by area of residence, 2004

Variable	UM	Urban	Rural	Urban/rural (%)
Total incomes	Euro/household	285.0	246.4	115.6
Incomes in cash	Euro/household	269.5	115.1	234
Consumption expenditures	Euro/household	207.9	157.3	132.1
Food consumption expenditures	Euro/household	186.0	108.9	170.7
Calories, out of which:	Kcal	2390	2613	91.5
of vegetable origin	%	77.4	76.9	100.7
of animal origin	%	22.6	23.1	97.8
Proteins, out of which:	Grams	77.5	83.8	92.5
animal	%	44.9	40.8	110.0
Cost of calorie	Euro/calorie	3.4	2.8	117.9

Source: Coordinates of the living standard in Romania incomes and consumption of the population in the year 2004, NIS, (2004).

in the peasant households (magiun – a traditional food, i.e. a sort of jam produced from plums without using sugar, or carnati de Plescoi – sausages made from mutton and beef meat). Table 9.5 presents the differences and characteristics of food consumption by the two main areas of residence.

By analysing the information in Table 9.5 one can state that, although the rural population has lower incomes (cash incomes in particular) than the urban population, it has higher food security, with an average daily consumption/capita of over 2600 calories compared with 2390 calories/day for the urban population. Nutritionists recommend an average consumption of about 2700 calories/capita/day for Romania. This reality is the effect of the self-sufficiency that characterizes most of the rural households, where 'access to food' is on a direct basis, with no intermediaries. At the same time, in the rural area the expenditures for services (heating, gas, running water, sewerage, and electricity) are much lower, as most households are not connected to many of these services.

The economic theory on consumers' behaviour associates a high share of food consumption expenditures (58.0 per cent in the rural area) with low incomes and poverty. In Romania's case, this is also an effect of the extremely high share of self-consumption in the food consumption expenditures of the rural population and of the extremely reduced share of income in cash. The traditional peasant households (subsistence and semi-subsistence) represent the great majority of rural households, having in general only sporadic connections to the input or agricultural markets. According to some estimates made on the basis of the results of the General Agricultural Census, only 92,000 individual holdings produce for the market (Figure 9.6).

According to the same estimates, out of the 4.4 million individual holdings, about 3.4 million can be considered as 'autarchic' (subsistence) households, while

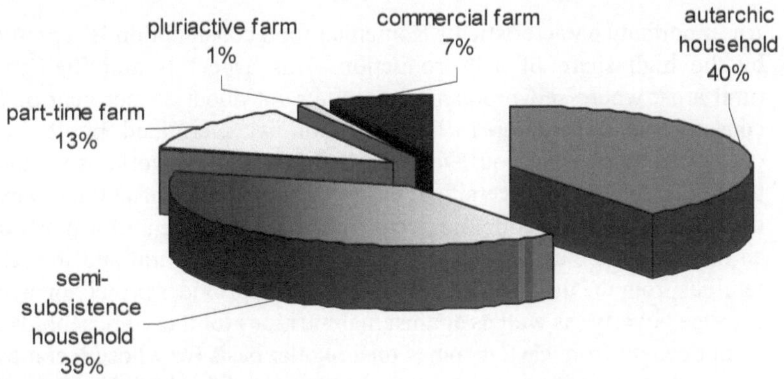

Figure 9.6 Type of individual agricultural holdings, by area

Source: Coordinates of the living standard in Romania incomes and consumption of the population in the year 2004, NIS, (2004).

about 1 million of individual households can be considered as of semi-subsistence type. Figure 9.6 shows that the 'autarchic' households use 40 per cent of the agricultural land in Romania, while semi-subsistence households use 39 per cent. In the rural area, where most of food comes from households' own production, a weaker correlation exists between the nutrition condition and incomes. Thus, although in rural areas incomes are lower than in towns, food consumption is higher than in urban areas. One can also observe the low cost of calorie in the rural population's food, which reflects a lower quality and a lower diversification of food consumption in the rural population.

9.5 Conclusions

The effects of transition upon the evolution of the consumption pattern evolution were mainly the increase of its subsistence and autarchic character, mostly for the rural population. The economic variables, represented by household incomes and expenditures, mainly reflect no cash availability and the poor connection to the market in the case of rural households. Thus, cash income represents only half of the total income of these households, while the other half is the value of food self-consumption.

The structure of food consumption expenditures is a proof of the autarchic character of rural households; it reveals that in this area, the solvent demand, represented by the quantity of purchased food, is much lower than in urban areas (only about half as much). However, in reality, in the rural area, there is a significant solvent food demand only for those products that cannot be obtained from people's own households (bread, sugar, oil, and certain alcoholic drinks) (Alexandri, 2000). To sum up, the following issues can also be observed:

1. An important characteristic of Romanian food consumption is represented by the high share of self-production. This aspect is amplified in the rural area, where self-production accounts for about 55 per cent of food consumption expenditures. This situation is determined by the weak connection of peasant households to the marketing network, as well as by the low employment diversification in the rural area. Agriculture is by far the most important economic sector in the rural areas, and it plays quite an important role as a food security source both for rural and for related families from the urban areas. Thus, agriculture provides protection against extreme poverty (as well as against malnutrition) for those households who do not benefit from cash incomes on a regular basis (or whose incomes are extremely low), i.e. peasant households, households with self-employed in non-agricultural activities, unemployed or farmer pensioners.
2. The period of economic growth, which started in the year 2000, resulted in an increase in incomes; this improvement has been noticed mainly in the urban area, the increase of incomes in the rural area being insignificant. This increase in incomes has led to a decrease in cereal consumption and an increase in meat and meat products consumption, starting with the year 2000; nevertheless, this situation is mainly a characteristic of the large urban centres.
3. Romania's consumption pattern reveals a duality. It is differentiated by the two main areas of residence i.e. urban and rural. The rural area features a high share of self-consumption and a lower diversification level, following the cycle of the seasons. Here, too, there are elements indicative of early modernity: these are found mainly in the population coming from the urban areas after 1990, as well as in the population who, after having left to work in foreign countries, are now returning to their native villages with money and with the standards of people who have seen other things, different from those that they have had at home so far.

References

Alexandri, C. (2000), *Securitate şi echilibru alimentar în România*, Editura Geea, Bucuresti.

Alexandros, N. (1995), *Agriculture mondiale – Horizon 2010*, Rome, Etude de la FAO.

Colman, D. and Trevor, Y. (2000). *Principles of agricultural economics: markets and prices in less developed countries*, Cambridge, England, Cambridge University Press.

Csaki, C. and Kray, H. (2005), *Romanian Agrifood Sector in a European Perspective*, World Bank, ECSSD, Working paper No.39.

Hertel, T., Brockmeier, M. and Swaminathan, P. (1997), *Sectoral and economy-wide analysis of integrating Central and Eastern European countries into the*

EU: Implications of alternative strategies, European Review of Agricultural Economics, Oxford, England, Oxford University Press.
National Institute of Statistics (1993–2005), *Bucharest Romanian Statistical Yearbooks*, Bucharest.
National Institute of Statistics (1998–2000), *Food balances*, Bucharest.
National Institute of Statistics (2002), Bucharest, Romania's Statistical Yearbook 2001–2003, Bucharest.
National Institute of Statistics (2004), *Coordinates of the living standard in Romania. Incomes and consumption of the population in the year 2004,* Bucharest.
National Institute of Statistics (2004), *Aspects regarding the life quality in Romania during the period 1999–2004,* Bucharest.
OECD (2000), *Review of Agricultural Policies*, Romania, Bucharest.
Padilla, M. (1996), *Les politiques alimentaires*, Paris, Editions Cujas.
Schmitz, M. (1997), *CAP and Food Security: Issues in agricultural competitiveness*, United Kingdom, Aldershot, Ashgate.
Senauer, B. and Roe, T. (1999) 'Food Security and the households', in: G. H. Peters and Joachim von Braun (eds), *Food Security, Diversification and Resource Management: Refocusing the Role of Agriculture?*, England, Ashgate, pp. 97–110.
Timmer, C. P., Falcon, W. P. and Pearson, S. R. (1983), *Food Policy Analysis*, The Johns Hopkins University Press.

Chapter 10
How Could Traditional Consumption Stimulate the Bakery Industry?

Iuliana Ionel

Introduction

In the early 1930s, Romania was a leading exporting country on the world grain market. However, following past policies and the restructuring of agriculture into a centralized sector, Romania ceased to be a supplier for the region and turned from being a net exporting country into a net importing country in the 1990s.

Grain markets in Romania are in a state of transition. The basic reforms are now completed: state controls have been removed, international and domestic trade in grain is unrestricted. However, the private sector storage and trading system is still in an early development stage: many storekeepers have only recently acquired their assets, typically through management-employee buy-outs, and the financing mechanisms for private traders and millers to acquire the crop are not yet fully functioning. International trading companies have entered the market and are actively seeking opportunities, mainly for external transactions in grain.

In Romania, over the last 10 years an average of 66 per cent of total arable land has been cultivated with cereals. In 2002, the main cultivated cereals were: corn and sorghum 47 per cent, wheat 41 per cent, barley 8 per cent, and the remaining 4 per cent were cultivated with oats and an insignificant area with rice.

Generally, agricultural producers in Romania cultivate cereals without having signed any contract with potential buyers of cereals. As a consequence, small producers take the decision to cultivate cereals depending on their own needs and also in general use various coping strategies so that after the harvest they might find a buyer for their cereals. If the price offered by the buyer is not attractive, individual producers prefer to store their cereals in their own household, even if the storage conditions may not be adequate. Cereals produced on typical agricultural holdings: agricultural associations or commercial farms, are partially sold after harvest if the farm concerned needs cash, or if partnership with a number of buyers can be established within a short time. However, a tendency is increasingly observed to build 'on farm storage' and to keep cereals within the farm, hoping to get a better price as no system of warehouse certificates has yet been implemented.

In recent years, mills or bakeries have ensured that an important quantity of cereals comes from their own farms, or many commercial farms have decided to build their own mill and bakery, so that they have their crop sold and minimize the risk to their trading. But there are also vertically-integrated units which produce, store and process cereals, either in the milling and baking units or in their own factories for concentrated fodder, destined to be used for their own livestock.

The milling and baking industry went through a restructuring process, implying that new small and middle capacity units appeared which are adapted to local demand. The factories wanted to ensure diversification of production and to improve the standard of in labelling and packaging of products, so that some of them received ISO certification. Still, further investments are needed in this sector in order to make it competitive, having in view Romania's accession to the European Union.

10.1 Marketing channels for grains

In general, cereals are cultivated in Romania without a firm contract concluded between agricultural producers and processors or traders of cereals.

As bread is a perishable product, relatively large in volume and small in value, with fast circulation and low possibilities of recycling returned quantities, its distribution represents the key point. Bread processing is carried out mainly during the night, and distribution takes place early in the morning starting at 4 a.m. The larger the distributor's area, the earlier is the distribution. Production is generally adapted to the distribution system and orders are also taken into account. This is why issues like how much is being bought, or what will be bought need to be known well in order to decide when to stop production. Because of the specific features in producing bread, the moment to stop activity has to be anticipated by three hours.

As a rule, bakers have their own transport means for the distribution of bread. Usually small bakers cover the local market, and big bakeries use more transport means to distribute bread and flour in bulk to big bakeries. There is a tendency to use specialized companies for the distribution of flour in sacks to small and medium-sized factories. The large bakers are increasingly using the specialized divisions of the companies specialized in distribution in order to handle flour for home consumption and for other bakery products that have a longer shelf life.

Generally, millers and bakers have their own shops but, in most cases, bread shops have been changed into general stores, as the daily circulation of bread is not high enough to justify the expense. Independent bakery stores do not allow the factories to plan their output accurately, as these stores are clients for many processors.

Supermarkets, as a rule, sell bakery specialties which have a longer shelf life (toast bread, deep-frozen products) and, in this case, deliveries take place twice a week. For snacks and biscuits, deliveries are made once a week.

There also are more modern modalities for selling bakery products. So, there are now stores working in franchising systems. In these shops, products are produced right in front of the clients on the basis of part-baked dough.

Table 10.1 Monthly consumption per capita (in kg)

Items	Total		Urban area		Rural area	
	2003	2004	2003	2004	2003	2004
Bread	9.55	9.14	9.33	9.29	10.23	8.92
Cornflour	0.98	1.79	0.68	1.03	1.9	2.83
Wheat flour	1.06	0.86	0.98	0.84	1.31	0.9

Source: Romania's Statistical Yearbook (2003 and 2004), National Institute for Statistics.

10.2 Traditional consumption

The Romanian milling and baking market is a market in which consumption has a decreasing trend (Table 10.1). The explanation for this can be also found in the history of other countries: bread consumption declines when the standard of living becomes higher, but at the same time the bread becomes more refined.

The modifications of consumption are already being felt in the Romanian baking industry: the number of active companies is down, while the stronger companies are consolidating their position.

Out of the 5,000 registered companies, 'probably 3,000 are not active on the market', while, of the remaining companies, only 30 are large companies, according to the Romanian Employers' Organization in the Milling, Baking and Pasta Industry (ROMPAN). The merging of companies continues, while the investments in modernization, marketing and distribution prevail compared with the investments in the enlargement of production capacities.

All these tactics target a single goal: the increase in the quality of products, and maintaining low prices so as to survive in an increasingly competitive market, estimated at about €1 billion. According to the statistics, the Romanians spent 31,200 billion lei for 2.5 million tons of bread in 2005, while 64 per cent of this quantity was consumed by people living in towns. A Romanian eats daily one loaf of bread that normally weighs 300–400 grams, while an American eats 85–160 grams of bread daily; 74.5 per cent of the town people prefer white bread. The main criteria taken into consideration when buying bread are the following: its taste (26.6 per cent), and the possibility to buy it from the nearest store (22.3 per cent).

About 130–140 thousand tons of bread are produced each month. An analysis of the bread production by months and years indicates a slightly decreasing trend from one year to another. The decrease in bread consumption may be the result of the increase in bread prices, accompanied by a decrease in the total population of Romania (about 100,000 per year).

Bakery expenditures represent about 25 per cent of the bread price (simple white bread). Romania imports several milling and bakery products, mainly from the European countries; imports reach about €9–10 million each year.

Table 10.2 Bread production, on a monthly basis, 2000–2003 (in thousands of tons)

Year	Jan.	Feb.	Mar.	Apr.	May	June	July	Aug.	Sept.	Oct.	Nov.	Dec.
2000	142.7	142.7	186.8	145.6	145.8	146.8	147.0	147.0	156.0	155.6	153.4	150.9
2001	149.7	148.5	153.8	135.1	139.1	138.3	138.9	138.9	136.7	137.8	134.9	132.3
2002	130.6	129.1	131.3	131.0	129.8	129.5	132.8	132.8	130.6	132.4	130.3	130.2
2003	128.4	128.1	131.4	131.4	132.7	132.2	134.5	133.8	133.6	135.6	313.2	130.7

Source: Romania's Statistical Yearbook (2003 and 2004), National Institute for Statistics.

Table 10.3 Imports of milling and bakery products, by countries of origin, 2002–2004 (in millions of euros)

	Years	Total	Austria	Belgium	France	Germany	Italy	Netherlands	Spain
Milling	2002	9.9	1.0	0.3	0.7	3.8	0.1	2.7	0.4
& baking	2003	9.0	0.3	0.2	1.0	2.5	0.6	2.7	1.1
products	2004	8.2	0.3	0.2	0.5	1.8	0.7	3.1	0.9

Source: Romania's Statistical Yearbook (2003 and 2004), National Institute for Statistics.

10.3 Investments and development

In recent years, the share of the milling and baking sector within the branch 'Industry of Food and Beverages' has stabilized at about 27–22 per cent, this evolution having a decreasing trend. Within the sector, the following sub-sectors stand out: baking and pastry industry (about 10 per cent), milling (about 5 per cent), biscuits (2.4 per cent). The main activity in the sector is the baking industry, whose value represents almost half of the milling and baking industry value.

In the period 1998–2002, the investments in the milling and baking sector accounted for 15–25 per cent of the net investments made in the industry of food and beverages.

The cumulated net investments in the period 1998–2002, made by the milling and baking companies, were about €390 million, of which 50 per cent were in the baking sector, 34 per cent in the milling sector, and 8 per cent in the biscuits sector. A maximum value was noticed in the year 2001, when the net investments totalled €146 million euros, of which 62 per cent were for the milling industry.

10.4 Diversification and competition: Two key studies

The *Bread Factory Dobrogea SA Constanta* was privatized in 1995. The shareholders of the company were the association of employees (96 per cent) and individuals who bought shares by public bid. A first step taken after privatization

Table 10.4 Share of the milling and baking industry within the industry of food and beverages (%)

	1998	1999	2000	2001	2002
Share of the milling and baking industry	36.5	23.2	26.9	24.6	22.1
Out of which:					
milling	4.9	5.0	9.2	5.9	5.4
baking and pastry	26.1	12.7	12.5	13.3	9.6
biscuits and cookies	1.6	1.8	2.0	1.4	2.4
pasta	0.4	0.3	0.3	0.4	0.5
other	3.5	3.3	2.8	3.6	4.3

Source: Romania's Statistical Yearbook (2003 and 2004), National Institute for Statistics.

was the restructuring of production and personnel, and investments worth $12 million were made.

The factory in Constanta has two mills with a total milling capacity of 520 tons/day. These mills were modernized first in 1994, then again in 2000, by purchasing advanced milling technology, with the Buhler trademark. The milling capacity of the factory is still higher, 610 tons/day: added to the 520 tons milled in Constanta are another 90 tons/day representing the production of the mill from Medgidia. At the same time, the factory in Constanta also includes a corn mill with a daily capacity of 50 tons. Of the produced flour, 40 per cent goes into bread production, the remaining quantity being sold separately. Dobrogea flour achieved a 20 per cent share on the Romanian market, the trading of this product having also the main weight in turnover. As regards the local bread market, Dobrogea has a share of 82 per cent, and the trade with this product represents 32 per cent in turnover. In total, Dobrogea has 21 production units, both in Constanta and in the rest of the county.

In order to keep its leading position on the domestic market, Dobrogea has also focused its attention on research activity, which is carried out within a separate division. Thus, in 1993, Dobrogea produced the first enzymatic ameliorator for baking 'Dan-Do-Pan' which enjoyed a great success among Romanian and foreign bakers. The ameliorator gives bread freshness, volume and aroma. Since 1999, Dobrogea has also produced flour correcting, which gives this product a rather homogeneous character. Last but not least, it should be noted that all the bread produced by Dobrogea is mineralized and enriched in vitamins, by which 40 per cent of the daily vitamin requirements for the human body is ensured, according to a recent study by Hoffman La Roche.

Since October 2000 Dobrogea SA has complied with ISO 9001 for the management of quality. Dobrogea made investments totalling almost €11 million, all in technology. In the Dobrogea mill €3.6 million were invested for the production line modernization. The fully automated manufacturing line for biscuits, producing 20 tons of biscuits per day, cost €4.2 million.

Another €3 million were invested in a production line for bread, delivering 30 tons of bread daily. The investment in the new line consolidates Dobrogea's position on the market and allows it to lower the price for this product. Dobrogea has 70 per cent of the bread sales and 60 per cent of the flour sales in the county of Constanta, as well as 22 per cent of the national flour market.

For the frozen baking and pastry products, Dobrogea has a market share of 80 per cent at the local level and 40 per cent at the national level. The company has 12 factories in the county of Constanta, 1.250 employees, and had a turnover of €72 million in 2005, up by 25 per cent compared with 2004.

Since the *Loulis Group* entered the Romanian market in 1999, the company has invested over €80 million. The company made its first profit only in 2005. However, the profit did not exceed 2 per cent of the turnover. In 2004, the company's investments focused on the modernization of the manufacturing line for sliced bread, on the implementation of Hazard Analysis of Critical Control Points (HACCP) norms on food hygiene and safety – imposed by the EU – and on building up a production line for frozen dough.

The company will continue to invest this year, too, but will direct funds of about €1.7 million to marketing and distribution activities. When entering the Romanian market, the Loulis Group took over the retail chain belonging to the Titan S.A. milling and baking unit. The company has thus diversified its production and distribution activities towards retail activities.

The Loulis Group management decided on a different approach for the bakeries, transforming them into a new identity, 'Family'. The image of the retail chain was related to the milling and baking activity of the company through the slogan 'Bread and more'.

In 2005, the company's management decided to rebrand the shop network. A stronger name was thus created through the association with Belforno, the brand Loulis holds on the baking segment, adapted, of course, to the specificity of the retail chain.

Fresh bakery products made by Loulis, be they traditional products or new assortments designed for the modern consumer, satisfy the most diverse preferences. These products are made of selected ingredients, using premium technology based on special recipes. The range of products includes:

- Bread: super loaf, extra loaf, mini diet, white bread, unsalted bread, home-made bread, traditional bread;
- Specialties: milk roll, extra roll, rye roll, fibro pan;
- Pastry with various fillings: cheese, vanilla, apples, cocoa, chicken;
- Starting in February 2005, Loulis S.A. is launching new types of fresh bread, under the Belforno Fresh brand: Whole grain bread, Black bread, and Extra Loaf.

In the second half of 2005, Loulis penetrated a new market, that of frozen dough. The company's management considered that its tradition in baking/pastry inherited from Titan S.A., complemented by solid knowledge acquired from the biggest company in the field in Greece, Katselis Sons (a shareholder of Loulis S.A.), represents a real advantage for activating on this market. The frozen dough production line, located in the industrial unit of Cernica, became operational last summer. At first, the products were addressed to the Loulis retail chain, under the Belforno Frozen brand. Starting in 2006, Loulis S.A. is targeting clients who would like to sell freshly-baked pastry products in their shops.

10.5 Conclusion

The small producers will be obliged to adjust to the new requirements, despite their low financial power. The small bakeries can only remain on the market if they meet the consumers' requirements, addressing a certain niche exclusively, and if they respect all the food safety rules. The small producers' best chance is to focus on the traditional and organic products, that is, on the value-added products.

The producers' opinion is that, at an overall level, the market will grow – not on a quantitative basis, but first by the consumers seeking quality products with value added, and then, by the increase in the bread price, as Romania joins the EU.

The processors in the milling and baking industry expect the following phenomena to appear after Romania joins the EU: 1) concentration in the milling industry; the number of companies on the market will reach only several hundreds in a couple of years; 2) the large industrial bakeries, the small ones, and those from the super/hypermarkets will coexist. However, their number will be significantly reduced, depending on the quality of products and the assurance of food safety; 3) bread consumption will decline over several years, from 113 kg/capita to 90–95 kg/capita; 4) only 10–20 per cent of the producers who are active at this moment will be able to continue their activity after Romania joins the EU; 5) the increase of prices after the accession will go hand in hand with the increase in industry profitability; 6) the quality of products is expected to rise faster than prices.

References

Fattouh, L. D. B. (2005), 'Concentration in the banking industry and economic growth', *Macroeconomic Dynamics*, vol. 9, n° 2, pp. 198–219.

Irwin, D. A. (2006), in *World Trade Review*, vol. 5, n° 3, Published online by Cambridge University Press, 19 Oct 2006, pp. 489–491.

Stiglitz, J. E. and Charlton, A. (2005), *Fair Trade for All: How Trade Can Promote Development*, New York, Oxford University Press.

Vikström, M. L. P. (2002), 'The determinants of structural change: Transformation pressure and structural change in Swedish manufacturing industry 1870–1993', *European Review of Economic History*, vol. 6, n° 1, pp. 87–110.

Chapter 11
Consumer Decision-Making with Regard to Organic Food Products

John Thøgersen

Introduction

Organic food production combines traditional and innovative food production methods with modern marketing principles. Organic production of foods is not the only, but it is the most far-reaching of currently applied measures to increase the sustainability of food production and consumption. As in other areas, there is a pronounced preference for voluntary measures to promote the choice of environmentally-friendly food products. The widespread subsidizing of organic farming (over and above the subsidies that farmers in general usually receive) influences consumers by increasing the supply and reducing the relative price of organic food products. In addition, there are state-supported labelling schemes to assist consumers who want to choose organic food products (e.g. Thøgersen, 2002). These voluntary measures seem to have had some success in promoting organic food products: at least, it is reported that the sale of these products is increasing in most European countries (see, e.g., Henson, 2002).

However, the switch from conventional to organic food proceeds rather slowly. The actual market shares of organic foods is considerably lower than what surveys of consumer preferences would predict, and this even in the Northern European countries where traditionally the population is relatively concerned about the environment (e.g. Grankvist, 2002; Magnusson et al., 2001; Wier et al., 2001). To some extent this gap between expressed consumer preferences and actual market shares can be explained by supply-side phenomena: insufficient supply and distribution, the premium price typically claimed for organic products; and their quality not always being as good as that of the conventional products. Another part of the explanation resides in the consumer, however. Even if practically all consumers would prefer products to be environmentally-friendly, not everyone weighs (i.e. values) the organic attribute equally highly (e.g. Thøgersen, 1997; Thøgersen and Bredahl, in press), nor does everyone equally strongly believe that the purchase of organic products makes any difference (e.g. Thøgersen, 2002), or trust the producer's claim that the product is actually organic (e.g. Bech-Larsen and Grunert, 2001).

Hence, although the market for organically-produced foods has been increasing, it remains relatively small both in Europe and in other parts of the World (Willer and Yussefi, 2004). If the organic market is to become substantial, more effective policy – and marketing – is needed. For this purpose we need a better understanding of consumer decision-making when choosing between organic and non-organic foods.

A major trend in the food market in Europe and elsewhere is the 'convenience' trend, that is, the increasing popularity of more or less pre-processed food (e.g. Scholderer and Grunert, 2005). However, consumers who need convenience in food, but also value organic food have received rather dismissive treatment by the organic food sector until now (e.g. Lockie et al., 2004). If the market for organic foods is to expand beyond that simply for fresh produce, it is of key importance to examine consumer views of processed organic foods and whether – or how – they diverge from their views of fresh produce.

This was the raison d'etre of the study reported in this chapter, which is part of the CONDOR project (Consumer Decision-making on Organic Products). The CONDOR project was the first to examine attitudes and behaviour in relation to both fresh and processed organic foods and to do so across a number of the EU Member States. It involved the development of a theoretically based consumer decision-making model for the purchase of organic food and the testing of this model in eight EU Member States.

The study's conceptualization of consumer decision-making is based on Ajzen's (1991) theory of planned behaviour (TPB), which was modified slightly for this purpose. According to this theory, consumer behaviour is co-determined by: (a) the individual's decisions as reflected in behavioural intentions; and (b) situational constraints and facilitators. Consumer intentions are co-determined by: (a) attitudes towards the behaviour; (b) perceived social pressure; and (c) perceived control or self-efficacy. Finally, these three constructs are based on the person's relevant beliefs and evaluations. We measured all these constructs in our study, with the limitation that we focused on only one type of consumer beliefs: beliefs about the consequences of buying organic food. We also measured consumers' abstract values, which are 'desirable goals, varying in importance, that serve as guiding principles in people's lives' (Schwartz and Sagiv, 1995, p. 93). Most important in the present context, consumers' value priorities are assumed to determine the importance of the expected consequences of their choices (e.g. Peter et al., 1999). Finally, we included consumer experience with, and subjective knowledge about, organic food as additional predictors in various phases of the decision- making process (Ellen, 1994; Moorman et al., 2004; Thøgersen, 1997).

We expected that consumer decision-making regarding organic food in different countries can be described by the same overall model, but that there would be differences in the strength of various relationships in the model due to different national conditions (e.g. differences in national food cultures, in organic agriculture's share of total agriculture, and in the maturity of the organic market), all of which translate into different constraints and facilitators for consumers wanting to buy organic food products.

For example, the market share of organic food, as a percentage of the total food market, varies a lot across Europe (Willer and Yussefi, 2004). Viewed in a dynamic perspective, this reflects that the diffusion of organic food (Rogers, 1995) has been far from synchronous across European countries, and that the national markets have reached very different levels of maturity (Kotler, 1991). This is illustrated in Figure 11.1 below, based on results from the CONDOR survey.

In the following, I present the basic model and summarize the main results from the CONDOR survey regarding consumer decision-making with regard to fresh and processed organic food across eight European countries.

11.1 Method

Two questionnaires were used to collect survey data in eight countries which differ substantially with regard to the market penetration of organic food: Denmark, Finland, Sweden, Germany, the UK, Italy, Spain, and Greece (Figure 11.1). One questionnaire focused on the purchase of fresh organic produce: tomatoes; and the other one on basically the same product, but in a processed form; organic tomato sauce. Except for referring to a different product, the two questionnaires were identical. Besides asking about self-reported buying frequency for the specific organic food product in question, the questionnaire contained operationalizations of the behavioural antecedents proposed by the TPB, in addition to subjective knowledge about organic food, basic human values, and experience with buying organic food as a general product category. There were also some questions not used for the present study. The questionnaires were developed in English and translated into the language of each country. In order to check the validity of the translations, questionnaires were back-translated into English.

11.1.1 Subjects

The research population consisted of individuals at least 18 years old and in charge of or sharing the responsibility for the household's grocery shopping. If the responsibility was shared, the person with the next birthday was asked to complete the questionnaire. In each of the eight countries, approximately 1000 respondents filled out questionnaires distributed by a professional market research company. Each country sample was randomly split into two sub-samples, each of about 500 people, and each receiving one of the two questionnaires. Apart from the requirement to deliver a representative sample, the specific recruitment procedure was left to the market research company and hence differed between countries (random in some, stratified random in other countries). Printed questionnaires were delivered to each respondent either by mail with a prepaid return envelope or by hand, in which case it was collected when the respondent had answered the questions. A demographic profile of each country sample is shown in Table 11.1.

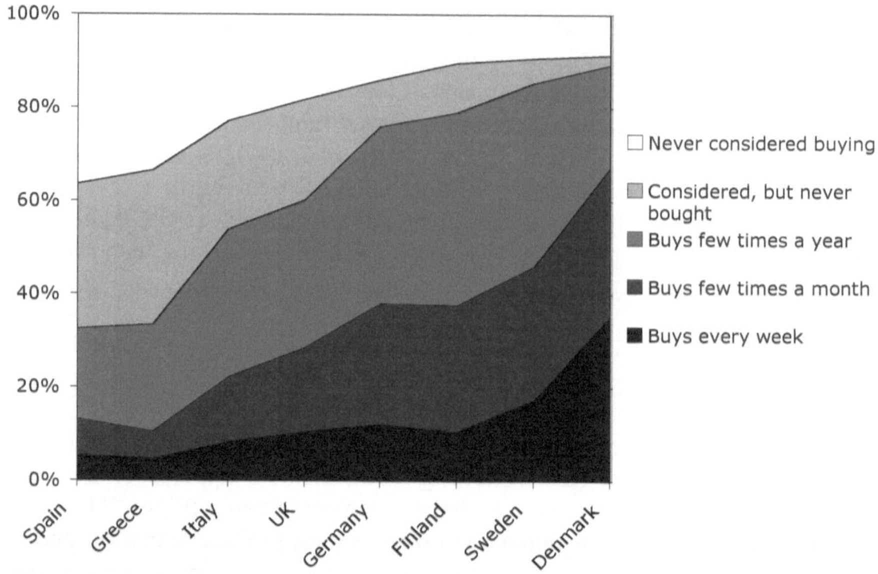

Figure 11.1 Consumer adoption of organic food in eight countries

Note: The market penetration of organic food was measured by means of the following question: 'From the following alternatives, please choose the one that best describes your shopping habits as regards organic food: 1) I have never bought, nor considered buying organic food,[1] 2) I have not yet bought, but I have considered buying organic food, 3) I buy organic foods few times a year, 4) I buy organic foods one or a couple of times a month, 5) I buy organic foods weekly.' In addition to these were two response alternatives which were coded as missing values in the following analyses: 'I always buy organic foods when possible' was coded as missing because it was realized ex post that it has an ambiguous time period reference and 'I have bought organic foods, but will not any more' was coded as missing because it would have made the scale non-monotonic. Overall, 8 per cent of the respondents chose one of these two options.

11.1.2 Variables

All responses were made on a 7-point scale unless stated otherwise. Items in the 'fresh tomatoes' questionnaires are presented below. The items in the 'tomato sauce' questionnaire are identical, except that when referring to a specific product the product is tomato sauce instead of fresh tomatoes.

Buying behaviour was measured by a single item only: 'How often, in the past 10 times when you bought fresh tomatoes, were these organic ones?' using a 5-point scale from 'never' to 'always'. Buying intentions were measured by two items: 'I intend to buy organic tomatoes instead of conventional ones in the near

1 Response alternatives were ordered, but not numbered in the questionnaire. Numbers indicate how responses were coded for the statistical analysis.

Table 11.1 Demographic profiles by country (%)

	ITA	GRC	SP	DK	UK	FIN	GER	SWE	All
Gender									
Male	43.7	28.2	14.2	24.9	23.6	30.7	48.3	38.7	31.6
Female	56.3	71.8	85.8	75.1	76.4	69.3	51.7	61.3	68.4
Age group									
18 – 30	19.8	19.9	18.5	9.9	21.4	21.9	21.4	17.8	18.7
31 – 45	38.9	38.5	33.3	33.0	33.7	24.5	28.9	26.6	32.0
46 – 65	35.8	33.3	33.1	41.6	35.3	42.8	33.1	36.8	36.6
over 65	5.5	8.2	15.0	15.5	9.6	10.8	16.6	18.8	12.7
Education Level									
< secondary school	8.6	24.7	56.2	36.9	7.0	13.5	21.6	21.6	27.8
secondary school	65.1	50.8	27.3	19.1	62.0	58.3	64.2	43.0	48.3
college/university	26.3	24.5	16.6	44.1	30.9	28.3	14.3	35.4	27.9
City area									
over 500 000	44.5	47.9	18.5	8.6	15.3	12.9	14.7	16.4	22.0
100 000 – 499 999	34.6	9.2	22.6	10.7	30.0	20.6	18.8	17.3	20.2
10 000 – 99 999	15.6	27.3	35.1	46.2	44.7	44.0	37.6	38.4	36.4
Less than 10 000	5.4	15.5	24.2	34.4	10.1	22.5	28.9	27.9	21.5
How many people are in your household (including yourself)?									
1 person	12.0	11.3	10.6	17.3	18.5	22.8	25.0	19.3	17.1
2 people	28.5	23.8	31.1	37.8	34.1	42.3	42.5	42.5	35.6
3 or more	59.6	64.9	58.3	45.0	47.4	34.9	32.5	38.3	47.3

future' (response categories ranging from 'definitely do not' to 'definitely do'), and 'I will buy organic tomatoes instead of conventional ones in the near future' (with response categories either 'extremely unlikely' or 'extremely likely'). The 2-item scale has acceptable construct reliability in all countries (range: .75 to .91).

Attitudes towards buying this particular product were measured by means of seven items, reflecting cognitive, affective, or moral evaluations. For instance, the cognitive attitude items were of the following semantic differentials: 'Buying organic fresh tomatoes instead of conventional ones is…' (harmful/beneficial, foolish/wise). Exploratory factor analysis was conducted for each country to see if the 7-items represented a single or several latent constructs. Using the Kaiser criterion, a single factor solution is produced in all eight cases.[2] The 7-item scale has excellent construct reliability in all countries (range: .89 to .92).

Perceived social norms to buy the organic product in question were measured with two items reflecting injunctive and descriptive norms, respectively: 'Most people who are important to me think that ('I should not buy' to 'I should buy')

2 In order to conserve space, this analysis is not shown. This and other omitted analyses can be acquired from the author.

organic tomatoes instead of conventional ones', and 'Most people whom I value would buy organic tomatoes instead of conventional ones' (response categories ranging from 'strongly disagree' or 'strongly agree'). The 2-item scale has a weak to acceptable construct reliability across countries (range: .55 to .73).

Perceived barriers for buying organic food products were measured by four items such as this: 'In general, for me to buy organic tomatoes instead of conventional ones would be …' (with response categories ranging from 'difficult' to 'easy'). In the following analyses, all items are coded so that a higher number indicates higher perceived barriers. The 4-item scale has acceptable construct reliability in all countries (range: .66 to .83).

Subjective knowledge (referred to as 'perceived uncertainty' in the following) was measured by means of a 5-item instrument developed by Flynn and Goldsmith (1999). A representative item is: 'When it comes to organic food, I really don't know a lot' (with response categories ranging from 'strongly disagree' to 'strongly agree'). In the following analyses, all items are coded so that a higher number indicates higher perceived uncertainty. The 5-item scale has acceptable construct reliability in all countries (range: .77 to .88).

The nine behavioural beliefs included in the survey are called 'modal salient beliefs' (Ajzen and Fishbein, 1980), and are based on interviews with a separate sample of individuals from the studied population. All items had the same format as this example: 'How likely is it that organic tomatoes are healthier than conventional ones?' (with response categories ranging from 'extremely unlikely' to 'extremely likely').

Experience with buying organic food was measured by the item: 'If you buy organic foods, please estimate for how many months or years you have been buying them' (a 6-point scale with response categories ranging from 'less than 3 months' to 'more than 5 years').

For measuring values, a shortened version of Schwartz's Portrait Value Questionnaire (PVQ) (Schwartz, in press; Schwartz et al., 2001) was used. The PVQ includes short verbal portraits of 40 imaginary persons. Each portrait describes a person's goals, aspirations, or wishes that point implicitly to the importance of a value. For each portrait, respondents are asked: 'How much like you is this person?' (on a 5-point graded scale with response categories ranging from 'not like me at all' to 'very much like me'). Our analysis of the data shows that among Schwartz's ten motivational domains, Universalism is the only or the dominant value guiding consumers' purchase of organic food in the eight countries (Thøgersen, 2006). Hence, basic values are represented by Universalism in this study. The 3-item Universalism scale has acceptable construct reliability in all countries (range: .69 to .74).

11.1.3 Statistical analysis

AMOS 5 (Arbuckle and Wothke, 1999) is used for the statistical analyses (confirmatory factor analysis [CFA] and structural equation modelling [SEM]). The main advantage of CFA and SEM is that it is possible to explicitly account for measurement error when a latent variable of interest is represented by multiple manifest variables. Measures of how well the implied variance-covariance matrix, based on the parameter estimates, reflects the observed sample variance-covariance matrix can be used to determine whether the hypothesized model gives an acceptable representation of the analysed data. AMOS was one of the first applications that offered Full Information Maximum Likelihood (FIML) to deal with item non-response. Non-response (full as well as item) reduces statistical power and may lead to biased parameter estimates. Recent years' extensive research into ways of dealing with missing data suggests that currently FIML is the most effective method to deal with missing data due to item-non-response, not only because it minimizes the loss of information and, hence, statistical power, but also because it leads to the most unbiased parameter estimates (Arbuckle, 1996), and this even in the case of non-normal data (Enders, 2001).

The usual assumptions about uncorrelated error terms and a simple structure factor pattern in the measurement model are applied. When there is only one item representing a latent construct, as for example with behaviour in the present study, the measurement error cannot be estimated but has to be set to a fixed value. In these cases, we fixed the error variance to zero.

Before conducting the analyses, the multi-item measurement instruments were checked for cross-cultural validity across the eight countries. Cross-cultural validity refers to the extent to which data collected by the same multi-item measurement instrument are comparable across different cultural environments (Bredahl, 2001). Measurement invariance was investigated using a procedure proposed by Steenkamp and Baumgartner (1998). They argue that measurement invariance is a matter of degree and suggest a stepwise procedure for revealing the level of measurement invariance.

In the present case, cross-cultural comparisons are limited to the strength of relationships between different constructs. In such cases, it is necessary to assume configural and at least partial metric (or scale) invariance (Steenkamp and Baumgartner, 1998). *Configural invariance* exists when the patterns of significant and nonsignificant factors are identical across countries. *Metric invariance* exists when factor loadings are the same across countries. At least one item per latent construct, in addition to the one fixed at unity to define the scale of each latent construct, needs to be metrically invariant in order for cross-national comparisons of structural model regression weights (path coefficients) to be meaningful (Steenkamp and Baumgartner, 1998). Steenkamp and Baumgartner (1998) recommend that invariance constraints should only be relaxed when it leads to highly significant improvements in model fit and that researchers evaluate not only chi-square differences, but also changes in alternative indices of model fit, especially those that take model parsimony into account, such as the root mean square of approximation (RMSEA).

Configural and metric invariance were tested by means of nested confirmatory factor analysis. The analysis showed that it is justified to assume partial metric invariance in this case.[3]

3 In order to save space, this analysis is not shown.

11.2 Results

Analyses (not shown in order to conserve space) revealed that it is admissible to apply the same model to both sub-samples, irrespective of whether the specific product is organic fresh tomatoes or organic tomato sauce. Hence, for the present purpose, the two samples are simply merged and a dummy variable taking the values 1 for organic tomato sauce and 0 for organic fresh tomatoes is included in analyses where it contributes significantly to explained variance in the dependent variable. Because of the complexity of the overall model, results are presented in three separate analyses. According to these analyses, the reasons and reasoning behind buying the organic tomato product are very similar across countries, but there are some – mostly minor – differences. Given that, when presenting comparative results from eight countries, the picture easily gets cluttered, in each step I first present results for Denmark, the country among the eight with the most mature organic market. After that I present the most important variations between countries.

11.2.1 Determinants of the attitude towards buying organic tomato products

Following the TPB, the attitude towards buying an organic tomato product is modelled as a function of the person's salient beliefs about the consequences of doing so. The person's general values (in this case, the priority given to Universalism) and experience with buying organic food in general are included as additional predictors. This model produces a satisfactory fit to the data and also satisfactorily explains variations in consumer attitudes towards buying organic tomato products (Figures 11.2 and 11.3).

It appears that attitudes depend primarily on beliefs about consequences, whereas basic values and past experience give more marginal direct contributions to explained variance. However, the significant correlations between, on the one hand, values, experience, and beliefs and, on the other, attitudes suggest that the latter two constructs have a stronger *total* (i.e. direct plus indirect) effect on attitudes than their regression coefficients indicate. For instance, both universalism and experience are positively correlated with the belief that organic food is good for your health ($r = .34$ in both cases). The implications of these indirect effects on the attitude will be discussed later.

The most important differences between countries concern the strength of relationships between beliefs about consequences and attitudes towards buying organic tomato products. Figure 11.3 shows unstandardized regression coefficients for these beliefs for each of the eight countries. Where it was statistically defensible,[4] regression coefficients were set equal across countries in order to remove unnecessary clutter. To the right of Figure 11.3 is the list of included

4 Here and in all other places, chi square difference tests with nested models are used to determine whether or not it is defensible to fix parameters to be equal. In order to save space, the specific test results are not reported, but they can be acquired from the author.

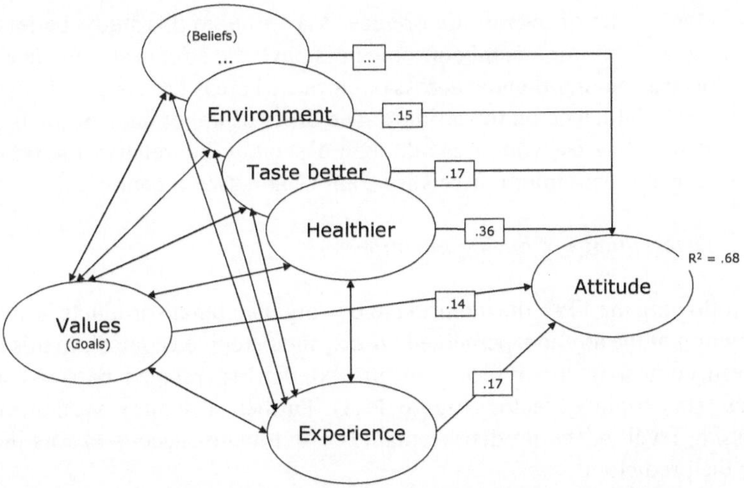

Figure 11.2 Determinants of the attitude towards buying organic food products, Denmark 2005

Notes: N = 1,113. Model fit: CFI = .98, RMSEA = .015. Standardized solution. In order not to clutter the picture, only the structural model is shown and only the three most influential beliefs. Correlations between Universalism, experience, and a number of beliefs, including the three most influential, are all statistically significant, $p < .05$.

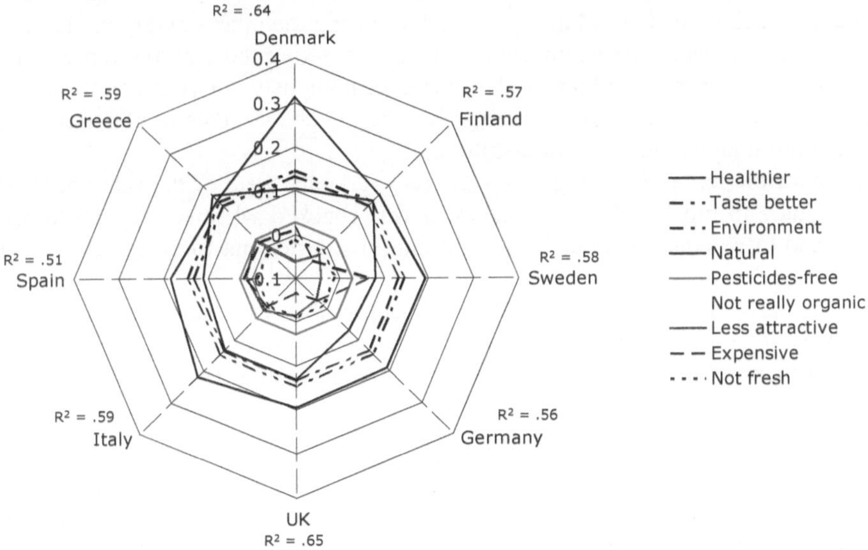

Figure 11.3 Belief importance in eight countries (unstandardized coefficients)

beliefs in their order of overall importance. According to this study, beliefs about health, taste and environmental consequences have the strongest influence on the attitude towards buying organic across countries whereas beliefs about costs have relatively little influence on the attitude. The order of importance of the beliefs is roughly identical across countries although especially the relative importance of naturalness and the premium price vary somewhat between countries.

11.2.2 Determinants of buying intentions

Again following the TPB, the intention to buy organic tomato products is modelled as a function of the attitude, perceived norms, and perceived control. In this model, perceived control is subdivided into an 'external' (perceived barriers) and an 'internal' (uncertainty) factor (Figure 11.4). Further, a dummy variable for the 'processing level' of the product (tomatoes = 0, tomato sauce = 1) was included among the predictors.

Pre-analyses (not reported) revealed that attitudes and perceived norms are strongly correlated (range: .71 to .94) in all countries, which means that there is a risk of multicolinearity. Inspection of the coefficients further revealed that the parameters estimated for three countries (UK, Sweden, and Greece) were indeed marred by multicolinearity: one of the coefficients (the attitude) had the wrong sign and the other one (perceived norms) was excessively large (Grewal et al., 2004). In cases of strong multicolinearity, it is logical that the involved independent variables must be more or less equally correlated with the dependent variable. Hence, it has been recommended to fix their regression weights to be equal in such cases (e.g. Marsh et al., 2004). Using the other country results as a yardstick, the parameter estimates resulting from fixing these regression weights seems to be reasonable, although the procedure obviously ignores the information that one (in this case, perceived norms) is a slightly better predictor of behavioural intentions than the other (i.e. the attitude).

Since the regression weights for the two variables were of the same order of magnitude in most of the other cases, it was checked whether they could also be fixed to be equal without loss of fit, and this turned out to be the case in all countries except Germany.

The pre-analyses revealed that it is admissible to assume a number of regression coefficients identical across groups of countries.[5] The fit indices indicate a satisfactory fit to the data and the model explains a satisfactory share of

---

5 The regression coefficient for the processing level dummy proved to be equal (and non-significant) in all countries except Denmark. The regression coefficient for uncertainty proved to be unequal in Denmark, Finland, Sweden, Germany, and the UK, and equal (and non-significant) in Italy, Spain, and Greece. The regression coefficient for perceived barriers proved to be unequal in Denmark, Germany, Italy, and Spain, and equal in Finland, Sweden, the UK, and Greece. The regression coefficient for perceived norms proved to be unequal in Denmark, Germany, Italy, and Spain and equal in Finland, Sweden, the UK, and Greece.

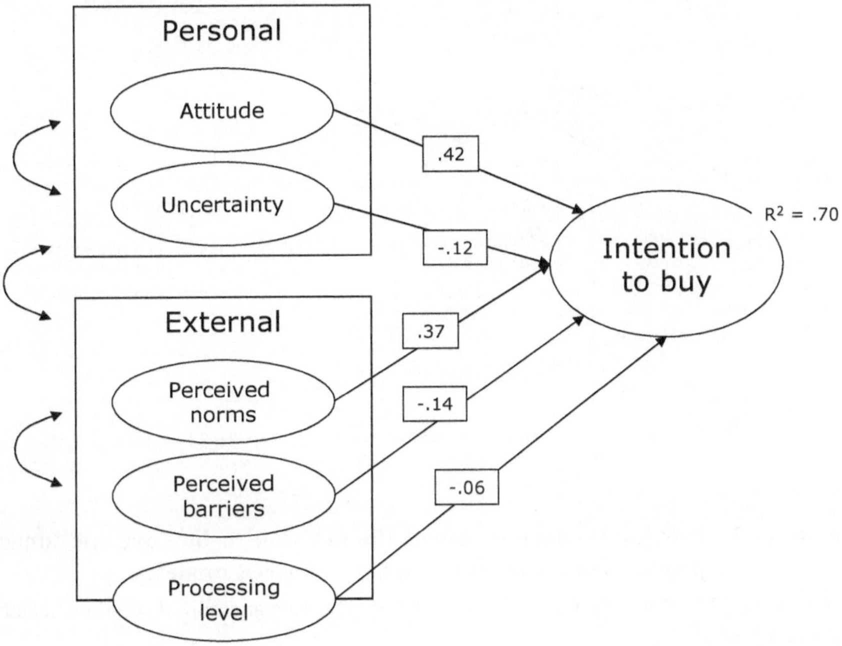

Figure 11.4 Proximal determinants of the decision to buy organic tomato products, Denmark 2005

Notes: Parameters from the eight-countries model, N = 8,113. Model fit: CFI = .93, RMSEA = .024. Standardized solution. So as not to clutter the picture, only the structural model is shown. Correlations between exogenous variables are statistically significant, except between perceived norms and processing level, $p < .05$.

the variation in intentions to buy organic tomato products in all countries (Figures 11.4 and 11.5).

The intention to buy organic tomato products apparently depends on external as well as personal factors. There is a strong relationship between the attitude and buying intentions in all countries. Hence, the more positive the attitude towards buying organic tomato products, the more likely it is that the consumer will decide to buy them (i.e. form a buying intention). However, except for Germany perceived social norms seem to play an equally big role for this decision. Hence, consumers who experience (or strongly believe) that their acquaintances buy organic food (i.e. hold strong descriptive norms about this behaviour) and/or that their acquaintances expect them to do so (i.e. hold strong injunctive norms: see Cialdini et al., 1990) are more likely to decide to buy these products.

The regression coefficient for the attitude proved to be unequal in Denmark, Germany, Italy, and Spain, and equal in Finland, Sweden, the UK, and Greece.

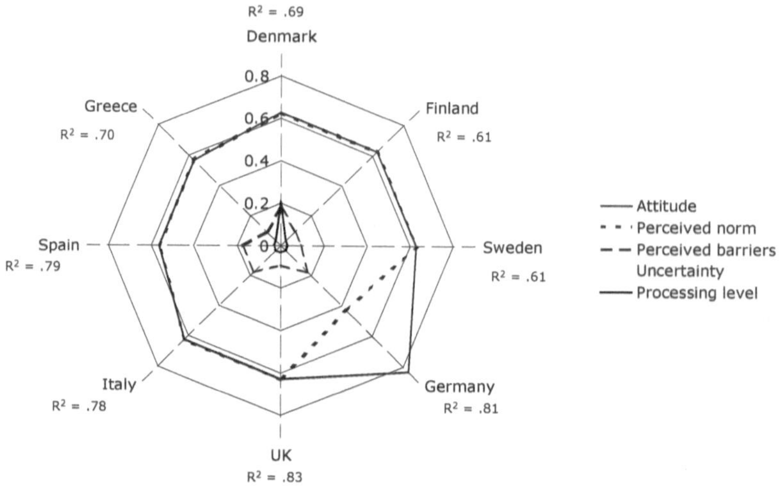

Figure 11.5 Proximal determinants of the decision to buy organic tomato products in the eight European countries model

Notes: N = 8,113. Model fit: CFI = .93, RMSEA = .024. Unstandardized solution, negative signs reversed.

Further, personal as well as external barriers seem to reduce the likelihood that a consumer will form a strong intention to buy organic food. If people feel uncertain about organic food they are less likely to make a decision to buy it, in spite of favourable attitudes and norms, and the same is true if they believe that the organic food products are difficult to obtain.

Figure 11.5 shows the results of the comparative analysis. Notice first that the processing level of the tomato product only has a direct effect on buying intentions in Denmark, but not in other countries. Some Danish consumers seem to reject organic tomato sauce for reasons that are not reflected in the four included predictors. The data material contains no clues as to what these reasons might be.

Besides this, the similarities across countries are much more striking than the differences. In all countries decisions to buy organic food products are co-determined – and with more or less equal weight – by personal and social considerations. Only in Germany are personal reasons significantly more important than social ones for these decisions. Perceived constraints and barriers also play a role at this stage in the decision-making in most of the countries, but seemingly a fairly marginal role.

11.2.3 Determinants of buying behaviour

Again following the TPB, buying behaviour is modelled as a function of buying intentions and constraints and barriers, including the person's uncertainty about

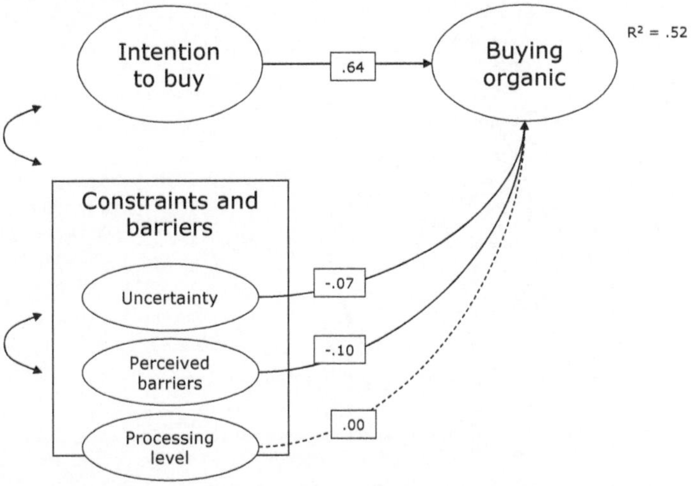

Figure 11.6 From intention to action, Denmark 2005

Notes: N = 1,113. Model fit: CFI = .94, RMSEA = .025. Standardized solution. So as not to clutter the picture, only the structural model is shown. All correlations between exogenous variables except between processing level and uncertainty are statistically significant, $p < .05$.

organic food. The dummy variable for 'processing level' was also included amongst the predictors at this level. This model produces a satisfactory fit to the data and also satisfactorily explains variations in consumer purchase behaviour. The results for Denmark are again discussed first (Figure 11.6).

In Denmark, there is a strong, positive relationship between buying intentions and (self-reported) buying organic tomato products. In other words, Danish consumers mainly seem to actually carry out decisions to buy organic food products. However, people who feel uncertain about organic food are less likely to buy it, in spite of favourable intentions, and the same is true for people who perceive that these organic food products are difficult to get hold of. In the Danish case, the 'processing level' dummy is not significant, which shows than any difference in buying behaviour between organic fresh tomatoes and organic tomato sauce is fully captured by buying intentions and perceived constraints and barriers.

The results of the comparative analysis are summarized in Figure 11.7. The most striking finding is that there seems to be a strong link between buying intentions and behaviour in the North of Europe, but not in the South. From Figure 11.1 we know that the organic markets in the South of Europe are less mature than in the North. This suggests that the reason for the weak intention-behaviour relationship in the South of Europe is low availability, which makes it difficult for organic consumers to act on their intentions (low 'actual control', see Ajzen, 1991). This inference may indeed be true, but then we have to conclude

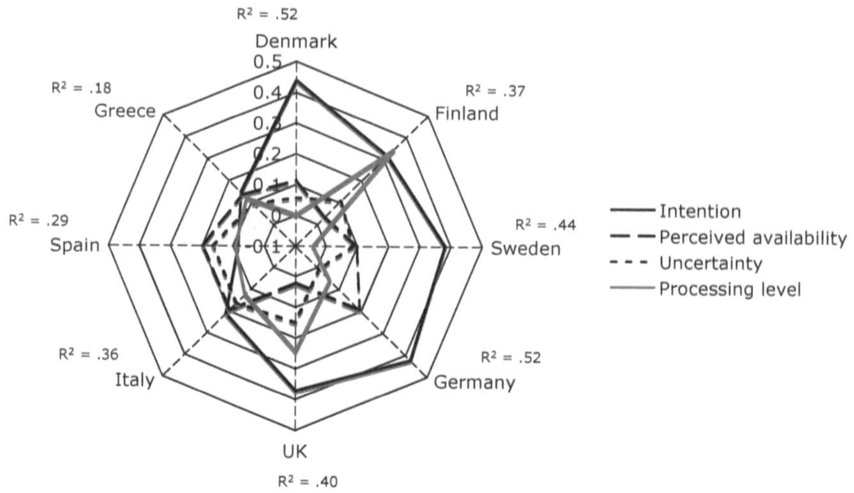

Figure 11.7 From intention to action in eight countries

Notes: N = 8,113. Model fit: CFI = .94, RMSEA = .026. Unstandardized solution, primarily negative signs reversed.

that individual consumer *perceptions* about barriers and constraints provide a very imperfect measure of the *actual* barriers and constraints on consumers wanting to buy organic food. The difference between North and South with regard to the intention-behaviour relationship is only reflected to a small extent in the finding that the influence of uncertainty and perceived barriers and constraints is stronger in the South.

The extent to which the processing level of the tomato product has a direct effect on buying behaviour also varies a lot across countries. This probably reflects a combination of two things: (a) that the actual supply situation and/or relative price for organic tomato sauce is much less favourable than for organic fresh tomatoes in some, but not in other countries; and (b) that consumers are often not consciously aware of this difference between organic tomato sauce and organic fresh tomatoes, which is why it is not (fully) reflected in the intentions and perceived barriers measures.

11.3 Conclusions and implications

The large sample size and cross-national coverage makes the present study rather unique. However, the study also has limitations, the most important of which are that it is a cross-sectional study, and that behaviour is measured by self-report only. The former limitation means that the evidence provided by this study is correlational only, and that inferences about causal relationships need to be based on information outside the study. Only to the extent that logic, experimental studies, or longitudinal studies provide the necessary evidence can we have confidence in the causal ordering proposed by the model. The second limitation means that the study may exaggerate the model's ability to predict real behaviour. Luckily, biased reporting is less of a problem for estimates of correlations than of means and, since only correlational data are analysed, biased reporting is hardly a serious threat to the conclusions of this study.

Hence, in spite of the limitations, a number of important conclusions about consumer decision-making with regard to organic food can be drawn: (a) overall, the reasons and reasoning behind consumer decision-making seem *not* to depend on the processing level of the food, although the actual purchase behaviour does; (b) consumer choices of organic food seem to be based on similar reasoning and reasons in the eight analysed countries, although there are (mostly minor) differences; (c) individual attitudes towards buying organic food are primarily based on beliefs about benefits (healthy, taste better, environmentally-friendly), while beliefs about costs or risks seem to be much less important; (d) social reasons seem to be as important as personal reasons for decisions about buying organic food in most countries; (e) there is uncertainty about what organic food is; and (f) perceived difficulties with purchasing organic food make some consumers give up, sometimes before, sometimes after having decided to buy organic food products. In the following, the policy implications of these conclusions for both governments and businesses are discussed.

The finding that the reasons and reasoning behind consumer decision-making do *not* depend on the processing level of the food implies that most organic consumers do *not* reject the idea of processed organic food. This is contrary to the widespread belief in the organic sector that organic consumers want only 'natural' organic products, that is, fresh produce. As a consequence of this belief, producers seem reluctant to develop convenience products for the organic market. This study suggests that it is time for producers to revise their opinions on this matter. Organic food consumers are no longer (if they ever were) just a small community of hard-core environmentalists and sympathisers. Increasingly 'mainstream' consumers with a busy lifestyle are in the market for organic food products (e.g. Latacz-Lohmann and Foster, 1997). Many mainstream consumers want sustainable food, but do not have the time to prepare it from scratch, at least not every day. Producers who supply good quality processed organic food products at a fair price should be confident that they will be rewarded by these consumers.

The finding that consumer choices of organic food are based on similar reasoning and reasons in the eight analysed countries indicates that there is a large unutilized market potential for organic food products (especially) in laggard countries (in the South of Europe, and probably elsewhere). It also suggests that, to a great extent, producers in laggard countries can base their marketing strategies on what has proved to be effective in countries with a more mature organic market. One successful strategy has been to take advantage of economies of scale and other sources of increased efficiency in a growing market to reduce the premium price charged for organic food products (e.g. Baecke et al., 2002). Since other studies have shown that the demand for organic food is actually quite price sensitive (e.g. Wier et al., 2003), reducing the premium price may be even more important in laggard countries with lower disposable incomes.

The analysis of attitude formation suggests that the benefits (healthy, taste better, environmentally-friendly) of buying organic food products are much more salient than costs or risks. This is lucky because it shows that the key differentiating attributes of organic food are highly valued by consumers. This study further reveals that attitudes towards organic food products are to a high extent value driven. Consumers with strong Universalism values hold more positive attitudes towards organic food irrespective of their factual beliefs (see also Dreezens et al., 2005), and they even hold more favourable *beliefs* about organic food. Schwartz (1994, p. 22) defines Universalism as 'understanding, appreciation, tolerance, and protection for the welfare of *all* people and for nature' [emphasis in original, JT]. The positive influence of Universalism is consistent with previous research on pro-environmental behaviour (e.g. Karp, 1996; Schwartz, in press; Thøgersen and Ölander, 2002), and it strongly suggests that consumers derive the self-relevance of buying organic food because it is considered a pro-social, pro-environmental type of behaviour (Thøgersen, 2006). The significant paths from Universalism to beliefs about organic food suggest that consumers with strong pro-environmental values tend to bolster their favourable beliefs about organic food by what has appropriately been called 'self-persuasion' (Aronson, 1999). This suggests that organic food producers need not 'cut corners' to promote their products to consumers. The documented environmental and ethical benefits of their products are strong selling points in the eyes of large consumer segments.

However, reservations are appropriate concerning the finding that beliefs about the costs and risks of buying organic are unimportant for consumer decision-making. As already mentioned, other studies have shown than the demand for organic food is actually quite price sensitive (e.g. Wier et al., 2003). What we *can* learn from this study is that the costs and risks of buying organic food products are not very salient in the interview situation. However, it seems reasonable to assume that costs such as a high premium price (and also risks such as a short shelf life) are much more salient in the shopping situation. This difference in situational salience may explain why survey studies tend to underestimate the importance of costs and risks for consumer choices.

The finding that social factors are as important as personal factors for decisions about buying organic food also has important policy implications. Consider the possible reasons why social influence is important in this area. Research suggests at least two: first, many consumers may need social proof that buying organic food is a sensible thing to do. Since organic food products usually command a premium price and since they are credence goods (Bech-Larsen and Grunert, 2001), many consumers may fear that they are being taken advantage of, and may doubt that buying organic food are indeed meaningful. But information that people like themselves, or people they respect, buy organic food may be helpful for overcoming this fear and doubt. Second, most consumers want to conform to what they perceive are reasonable expectations about proper behaviour (Bicchieri, 2006). The finding that positive attitudes towards buying organic food are to a great extent derived from Universalism values suggests that this is also relevant for organic food purchases. It seems unlikely that people with strong Universalism values should feel that *they* are the only ones who should buy organic food. Most likely, they expect other people like themselves to do the same, and also that this expectation is reversed. Hence, it is probable that both descriptive and injunctive norms can be mustered in support of promoting organic food products. Both social descriptive information and norm-based appeals may be effective means in this connection.

It was found in this study that not only social, but also factual uncertainty dampens the demand for organic food. Many consumers in the countries covered are uncertain about organic principles, the control system, and/or other key characteristics of the organic food production system. From this it follows that consumer education and information is still an indispensable component of policy and marketing in this area.

Finally, the negative influence of perceived (external) barriers that was found also has important policy implications. This finding reflects (although in an imperfect way) that organic food is still less available than its conventional competitors, even in the most mature organic markets. The importance of availability for consumer choice of organic food products has previously been documented by the experience from Denmark and other, more mature, organic markets (e.g. Michelsen, 1993; O'Donovan and McCarthy, 2002). This finding not only emphasizes the importance of an effective distribution system but also that a good supply of organic food is able to create its own demand.

In sum, there are many possibilities for improving the promotion of organic food products. Governments looking for ways to improve the sustainability of the food system have until now focused on subsidizing organic agriculture, sanctioning third-party certification, labelling, and a control system, supporting information campaigns, and – in some countries – including organic food in their 'green procurement' policy for governmental institutions. All of these initiatives are worthwhile and should be continued and even intensified in most countries. The most import contributions of governments are measures to create more favourable prices for, and increased trust in, organic food products. Companies in the organic sector should continue to supply organic products of high quality and at a fair price. Processors in the sector play an important role in this connection, especially for seizing opportunities for developing products with an increased level of convenience for the increasingly time-starved mainstream consumers. Finally, producers and processors should work on intensifying the distribution of organic food products both through conventional retailers and through special distribution channels and they should utilize the many pieces of advice for effective marketing communication given in the mushrooming research literature in this area.

Acknowledgements

The CONDOR project was carried out with financial support for the European Communities, under the 5th Framework Programme, project no. QLK1-2002-02446. For more information, visit http://www.condor-organic.org/. I am indebted to my partners and collaborators in the CONDOR project, who all deserve credit for the insights conveyed in this chapter. I alone am, of course, responsible for any errors, misunderstandings, or unclear formulations.

References

Ajzen, I. (1991), 'The theory of planned behavior', *Organizational Behavior and Decision Processes*, vol. 50, pp. 179–211.

Ajzen, I. and Fishbein, M. (1980), *Understanding attitudes and predicting social behaviour*, Englewood Cliffs, Prentice-Hall.

Arbuckle, J. L. (1996), 'Full information estimation in the presence of incomplete data', in: G. A. Marcoulides and R. E. Schumacker (eds.), *Advanced structural equation modeling: Issues and techniques* , Mahwah, N.J., Lawrence Erlbaum Associates, pp. 243–277.

Arbuckle, J. L. and Wothke, W. (1999), Amos 4.0 users' guide. Chicago, Ill., Marketing Division SPSS Inc., SmallWaters Corporation.

Aronson, E. (1999), 'The power of self-persuasion', *American Psychologist*, vol. 54, pp. 875–884.

Baecke, E., Rogiers, G., Cock, L. D. and Huylenbroeck, G. V. (2002), 'The supply chain and conversion to organic farming in Belgium or the story of the egg and the chicken', *British Food Journal*, vol. 104, n°s 3/4/5, pp. 163–174.

Bech-Larsen, T. and Grunert, K. G. (2001), 'Konsumentscheidungen bei vertrauenseigenschaften: Eine untersuchung am beispiel des kaufes von ökologischen lebensmitteln in deutschland und dänemark', *Marketing ZFP*, vol. 23, pp. 188–197.

Bicchieri, C. (2006), *The grammar of society. The nature and dynamics of social norms*, Cambridge, Cambridge University Press.

Bredahl, L. (2001), 'Determinants of consumer attitudes and purchase intentions with regard to genetically modified foods – results of a cross-national survey', *Journal of Consumer Policy*, vol. 24, n° 1, pp. 23–61.

Cialdini, R. B., Reno, R. R. and Kallgren, C. A. (1990), 'A focus theory of normative conduct: Recycling the concept of norms to reduce littering in public places', *Journal of Personality and Social Psychology*, vol. 58, pp. 1015–1026.

Dreezens, E., Martijn, C., Tenbult, P., Kok, G. and de Vries, N. K. (2005), 'Food and values: An examination of values underlying attitudes toward genetically modified and organically grown food products', *Appetite*, vol. 44, pp. 115–122.

Ellen, P. S. (1994), 'Do we know what we need to know? Objective and subjective knowledge effects on pro-ecological behaviors', *Journal of Business Research*, vol. 30, pp. 43–52.

Enders, C. K. (2001), 'The impact of nonnormality on full information maximum-likelihood estimation for structural equation models with missing data', *Psychological Methods*, vol. 6, pp. 352–370.

Flynn, L. R. and Goldsmith, R. E. (1999), 'A short, reliable measure of subjective knowledge', *Journal of Business Research,* vol. 46, pp. 57–66.

Grankvist, G. (2002), 'Determinants of choice of eco-labeled products', Unpublished Doctoral dissertation, Göteborg University, Göteborg.

Grewal, R., Cote, J. A. and Baumgartner, H. (2004), 'Multicollinearity and measurement error in structural equation models: Implications for theory testing', *Marketing Science*, vol 23, n° 4, pp. 519–529.

Henson, S. (ed.) (2002), The market for organic food products in Europe (Special Issue), *British Food Journal*, vol. 104, n°s. 3/4/5, pp. 103–406.

Karp, D. G. (1996), 'Values and their effect on pro-environmental behavior' *Environment and Behavior*, vol. 28, pp. 111–133.

Kotler, P. (1991), *Marketing management. Analysis, planning, implementation, and control* (7th ed.), Englewood Cliffs, NJ, Prentice-Hall.

Latacz-Lohmann, U. and Foster, C. (1997), 'Form "niche" to "mainstream" – strategies for marketing organic food in germany and the UK', *British Food Journal*, vol. 99, pp. 275–282.

Lockie, S., Lyons, K., Lawrence, G. and Grice, J. (2004), 'Choosing organics: A path of analysis factors underlying the selection of organic food among Australian consumers', *Appetite*, vol. 43, n° 2, pp. 135–146.

Magnusson, M. K., Arvola, A., Hursti, U.-K. K., Åberg, L. and Sjöden, P.-O. (2001), 'Attitudes towards organic foods among Swedish consumers', *British Food Journal*, vol. 103, n° 2, pp. 209–226.

Marsh, H. W., Hau, K.-T. and Wen, Z. (2004), 'Structural equation models of latent interactions: Evaluation of alternative estimation strategies and indicator construction', *Psychological Methods*, vol. 9, n° 3, pp. 275–300.

Michelsen, J. (1993), 'MarkEdsføring af økologiske produkter', Esbjerg, Sydjysk Universitetscenter.

Moorman, C., Diehl, K., Brinberg, D. and Kidwell, B. (2004), 'Subjective knowledge, search locations, and consumer choice', *Journal of Consumer Research*, vol. 31, pp. 673–680.

O'Donovan, P. and McCarthy, M. (2002), 'Irish consumer preference for organic meat', *British Food Journal*, vol. 104, n°s 3/4/5, pp. 353–370.

Peter, J. P., Olson, J. C. and Grunert, K. G. (1999), *Consumer behavior and marketing strategy,* European edition, London, McGraw Hill.

Rogers, E. M. (1995), *Diffusion of innovations*, 4th ed. New York, Free Press.

Scholderer, J. and Grunert, K. G. (2005), 'Consumers, food and convenience: The long way from resource constraints to actual consumption patterns', *Journal of Economic Psychology*, vol. 26, n° 1, pp. 105–128.

Schwartz, S. H. (1994), 'Are there universal aspects in the structure and content of human values?', *Journal of Social Issues*, vol. 50, n° 4, pp.19–45.

Schwartz, S. H. (in press), 'Robustness and fruitfulness of a theory of universals in individual human values', in: A. Tamayo and J. Porto (eds.), *Valores e trabalho* [values and work], Brasilia, Editora Universidade de Brasilia.

Schwartz, S. H., Melech, G., Lehmann, A., Burgess, S., Harris, M. and Owens, V. (2001), 'Extending the cross-cultural validity of the theory of basic human values with a different method of measurement', *Journal of Cross-Cultural Psychology*, vol. 32, n° 5, pp. 519–542.

Schwartz, S. H. and Sagiv, L. (1995), 'Identifying culture-specifics in the content and structure of values', *Journal of Cross-Cultural Psychology*, vol. 26, n° 1, pp. 92–116.

Steenkamp, J.-B. E. M. and Baumgartner, H. (1998), 'Assessing measurement invariance in cross-national consumer research', *Journal of Consumer Research*, vol. 25, pp. 78–90.

Thøgersen, J. (1997), 'Understanding behaviours with mixed motives. An application of a modified theory of reasoned action on consumer purchase of organic food products', Paper presented at the The Association for Consumer Research, European Conference, 25–28 July, 1997, Stockholm.

Thøgersen, J. (2002), 'Promoting green consumer behavior with eco-labels', in: T. Dietz and P. Stern (eds.), *New tools for environmental protection: Education, information, and voluntary measures*, Washington DC, National Academy Press, pp. 83–104.

Thøgersen, J. (2006), 'Are consumers guided by selfish or unselfish motives when they buy organic food?', submitted for review.

Thøgersen, J. and Bredahl, L. (in press), 'Cross-national and lifestyle differences in consumer choice criteria and motives with regard to organic food', *Asia-Pacific Advances in Consumer Research*, vol. 7.

Thøgersen, J. and Ölander, F. (2002), 'Human values and the emergence of a sustainable consumption pattern: A panel study', *Journal of Economic Psychology*, vol. 23, pp. 605–630.

Wier, M., Hansen, L. G., Andersen, L. M. and Millock, K. (2003), 'Consumer preferences for organic foods', in: OECD (ed.), *Organic agriculture – sustainability, markets and policies*, CABI Publishing, pp. 257–272.

Wier, M., Hansen, L. G. and Smed, S. (2001), 'Explaining demand for organic foods', paper presented at the 11th annual EAERE Conference, June 2001, Southampton.

Willer, H. and Yussefi, M. (eds.) (2004), *The world of organic agriculture. Statistics and emerging trends 2004*, Bonn, International Federation of Organic Agriculture Movements.

PART III
Mass and Segmentation in Traditional Food Markets

Chapter 12

Local Honey Production: Export or Indigenous Growth?

Teresa de Noronha Vaz

Introduction

Although there is almost no fossil proof, bees probably developed when flowering plants appeared, about 146 to 74 million years ago. The oldest known fossil bee dates from 96 to 74 million years ago and was found in New Jersey, USA. The precursor of the honeybees may have been living about this time, but fossils of the true Apis type were first discovered in the period of 22 to 25 million years ago, in Western Germany, and scientists guess that bees probably progressed from hunting wasps that survived on a vegetarian diet based on nectar.

Bees of all kinds belong to the order of insects known as *Hymenopter*, which comprises about 100,000 species. Of the around 25,000 described species of bees, the majority are solitary bees, laying their eggs in self-excavated tunnels. The females bees play a very special role: in a few cases females may share a single tunnel system, in others there may even be a semi/social organization with a hierarchy.

Honeybees belong to the family of social bees. These nest in colonies led by a single fertile female, which is generally the only egg layer in the colony. The other females are in charge of foraging for nectar and other tasks such as feeding the queen and the larvae and cleaning. Honey and pollen is stored, and larvae are raised in cells made from wax secreted by the worker bees. This social grouping varies in number, depending on the species of the bees. Some grouping may comprise just a few dozen insects, and may be annual, as in the case of bumblebee colonies, where others, like honeybees, may number several tens of thousands and continue for several years (Milner, 1996).

Two attributes of honeybees, which have been essential to their evolution and biology, are their clustering behaviour, as well as their ability to cool the nest by evaporation of water collected outside. Thus, colonies are able to control temperature within the nest, irrespective of the external temperature. The genus Apis was thus able to colonize a wide variety of environments, ranging from tropical to cool temperate. The Meliponinae, which lack this capability, are confined to tropical regions. Another behavioural characteristic of honeybees is the communication of information regarding food sources and the recruitment of foragers using 'dance language'. This exact dissemination of information concerning the direction and distance of forage areas allows a very efficient use of the food sources. Whereas most types of bee were indigenous to all the continents, bees belonging to the genus Apis (honeybee) originated in Asia, Africa and Europe.

It is difficult to agree on the period when man started to raise bees. Cave paintings in Spain, dating from 7000BC, show the earliest records of beekeeping. The earliest record of keeping bees in hives was found in the sun temple erected in 2400BC near Cairo. The oldest beehive ever found was located in Central Europe and dated 2000 years old. It was built of wicker in a campanula form and covered with clay and dung. Recently, another old beehive was found, close to Oldenburg, in Germany. This time, the find was not more than 500 years old and was still in a good condition, containing its initial cell structure and respective bees. Similarly to the present models, its form was shaped from an opened tree trunk, allowing the bees' free entrance and exit. These discoveries represent the two ancient forms to produce beehives – of lintel, if they were placed in the plains, or in wood, if they were located nearby forested areas.

Apiculture has been practised in Europe and Asia throughout recorded history. In the American regions, where the true honeybees were originally absent, the early settlers imported the bees with which they were familiar. Thus, Iberian bees were taken to Brazil and North European bees to North America, Australia and New Zealand. Whereas the Iberian bees were unsuited to the tropical climate of South America and failed to establish a feral population, the North European bees adapted well to the harsher conditions, and feral colonies quickly established themselves over a wide area (Milner, 1996). It was formerly believed, not only by ordinary beekeepers but by some recognized scientists, that improvements in the desirable attributes of honeybees, productivity, docility, resistance to disease, for example, could be achieved by cross-breeding different races. However, it is now widely accepted that the best way to improve bee stocks is by selective breeding within a single subspecies.

Indeed, in most parts of the World, especially where beekeeping is practised on a commercial scale, the Italian bee has proved the most popular: not aggressive, rapidly procreating, ready to continuously multiply providing food is available, it is the most suitable species for those countries in which Spring has a late start. When the nectar flows discontinuously and in bad weather conditions, feeding may be necessary during the sterile periods, and also in Spring and Autumn. For some species that cannot survive hard winters, the migratory beekeeping has been adopted; new colonies are raised each spring in the southern American states and transported to the forage grounds of the North.

Man has made multiple uses of the bee's products for many centuries: the honey was the only source of sugar available in abundant amounts; the wax was the first, or one of the first, plastic substances; finally, the propolis has been an important antiseptic in the popular medicine (Krell, 1996).

The ancient Egyptians used honey as a sweetener, as a gift to their Gods and even as an ingredient in the embalming fluid. Honey cakes were baked by the Egyptians and used as an offering to placate the Gods. The Greeks, too, made honey cakes and offered them to the Gods. They viewed honey as not only an important food, but also as a healing medicine. Greek recipe books were, and still are, full of sweetmeats and cakes made from honey. Cheeses were mixed with honey to make cheesecakes, described by Euripides in the fifth century BC as being 'steeped most thoroughly in the rich honey of the golden bee'. The Romans also used honey as a gift to the Gods and they also used it extensively in cooking. Beekeeping flourished throughout the Roman Empire. Once Christianity was established, honey and bees wax production increased greatly to meet the demand for church candles. Honey continued to be of importance in Europe until the Renaissance, with the arrival of sugar. Today the multiple applications and studies resulting from beekeeping are immense. We recommend a visit to a gathering exercise by the Michigan Tech University to evaluate the practicalities of this branch of activity (School of Forest Resources and Environmental Science, MTU, 2006)

In spite of the fact that the traditional consumption has taken good notice, and use, of all those products, since new industrial production developed and urbanization increased, only honey was able to easily integrate in the international marketing channels, and the other products have for many years been mainly oriented towards irregular local markets.

12.1 Honey, other bee by-products and their respective uses

12.1.1 The honey

Honey is a completely natural food which results from a saturated solution of sugars, mainly fructose and acid glucose, enzymes, minerals and some aromatic substances. It is produced by the bees once they have been nourished by the flowers' nectars. It is this nectar that provides the honey with its specific perfume and flavour. More precisely, the set of botanical varieties feeding the bees form the bases for the particular specific quality of each type of honey.

The natural process is simple. Bees produce honey as food stores for the hive during the winter months, when flowers are not in bloom, and therefore there is little nectar available. The nectar collected by the honey bees from flowers and plants is carried to the hive or nest and is then passed on to the worker bees, who prepare it for storing, by adding enzymes. As the nectar is transferred to the wax storage chambers, water is evaporated, and it is this process, combined with the enzyme activity that converts the nectar into honey. But, a hive only needs 20–30 lb of honey to survive an average winter, which means that the extra honey can be harvested. A strong colony can produce 2–3 times more honey than they need.

Bees can fly up to six miles, although one or two is more common. Within this radius, depending on the flower variety, they can gather nectar from many different types of flower (multifloral/polyfloral honeys, Chattopadhyay et al., 1996) or from one kind of flower (unifloral/monofloral honeys, Ruoff et al., 2004).

Besides the classification by floral source, honeys can also be described according to geographical origin, from a particular country or region. This is because a honey of the same flower type will still vary from country to country, because of difference in climate and soil. Altogether, we can classify the different honey types as: 1) Monofloral, when it is made basically from a single floral source, such as Acacia, Pine, Orange Blossom, Lime, Rosemary, Thyme, Sunflower, Clover, Leatherwood, Eucalyptus (Oddo, 2004); 2) Polyfloral/ Multifloral, when it is made from the nectar of many different flowers; or 3) Blends, if resulting from combinations of different types of honey, combined to achieve a particular taste, just like a tea or a whisky. In fact, if we start considering the production of monofloral types of honey, we should rely on the fact that it is produced by extensive culture, most probably with constant productivities and qualities that can be easily worked out, therefore achieving high quality standards.

In any case, honey is a complex mixture, consisting of 80 per cent natural sugars, 18 per cent water and 2 per cent minerals, vitamins, pollen and proteins. Of honey's 80 per cent natural sugar content, approximately 70 per cent consists of fructose and glucose. The proportion of these two sugars determines whether a honey is clear or set. Both types of honey are equally pure and additive free. The higher the fructose content, the longer the honey will remain as a liquid. Some honeys contain higher glucose content than fructose and, therefore, are very likely to crystallise swiftly. There is no difference in the taste or nutritional value of these two states. The extraction process – which corresponds to the removal of honey from the comb – is essentially the same for all types and grades of honey. All the honey is slightly heated, to melt naturally occurring crystals and to facilitate the filtering out of very small organic impurities from the hive, e.g. wax.

In order to offer the consumer a product that keeps all the characteristics it had when it was deposited by the bees in the honeycomb cells, the techniques applied during the production process must include the necessary precautions, namely: 1) extraction of the honey from the honeycombs, using a centrifugation process; 2) purification, performed by means of filtration and decanting; 3) bottling without thermal intervention, since under our climatic conditions the honey is liquid at the moment of extraction.

The first phase of the honey-making process is called uncapping, which consists of eliminating the opercula that close the cells containing the honey. For this, previously heated knives or machines are used to perform the uncapping. The extraction takes place afterwards, by means of centrifugal extractors that allow the honeycombs to be reused. The extraction is followed by a purification process, to eliminate the air bubbles, the wax particles, and any other impurity found in the honey during the extraction. It can be performed through decanting or filtering. Very often, the honey purification is made by filtering and decanting successively. At this stage, the honey is ready to be bottled. The pasteurization is performed in the companies where the honey is industrially prepared. The technique generally used to minimize the damage to the product involves the application of high temperatures (78° C) for a few minutes. The result obtained is a honey protected from the risks of fermentation, once the leavens that may be present are destroyed, and that honey will remain liquid for some months, that is, until the foreseen consumption time.

Many of the uses of honey are associated with its therapeutic properties, although frequently it is the monofloral honeys that are considered to have the pharmacological attributes (Titěra et al., 2004). Moreover, an anti-bacterial activity was detected both in the honey and in the diluted solutions. However, it is better to consider the honey as a food rather than as a medicine/pharmaceutical. The honey composition, which comprises glucose and fructose associated with organic acids, mineral salts, enzymes, aromas and many other substances, makes it a unique and very particular food. The honey is glucose of high energy. Since it is composed mainly of simple sugars, it is easily digestible. It enters directly into the circulation and is therefore immediately used, while the fructose is consumed more slowly and functions as an energy reserve, since before being used by the organism, it must be transformed into glucose. Therefore, honey provides an immediate source of energy, without requiring a digestive process.

In addition, in recent years, honey has been added to powdered milk and to baby food in general, since it improves the tolerance to cow's milk. Honey can always replace the sugar. Either as a liquid or crystallized, monofloral or multifloral, of a delicate or an intense flavour, sweet, floral, fruited, herbaceous, piquant, bitter, it can be chosen according to each person's taste. It can be consumed as it is for breakfast and for tea, and as a sweetener in drinks. It has many uses in cooking, e.g. as a glaze for hams and in salad dressings and as an ingredient in the preparation of sweets that do not need baking or boiling – for example, many regional traditional sweets that include honey as an ingredient.

Finally, we cannot forget that, besides being used in the food sector, honey has always been largely used in other areas (Wanderley et al., 2004), mainly in cosmetics.

12.1.2 The pollen

After nectar gathering, pollen gathering is the bees' main activity. In fact, this is the only source of proteins, lipids and mineral salts for the hive. Throughout the years, man has always used pollen for its nutritional and medicinal qualities. The use of this product in the cosmetics industry is a recent practice that dates back just from the second half of the 20th century (Bryant, 2001).

The yearly collection of this product in a hive can vary between 20 and 30 kg, sometimes reaching 50 kg. The collection is not done regularly, since it varies with the season, being more intense in the Spring-Summer / Summer-Autumn.

The pollen found in the honeycomb cells is solid, and it cannot be gathered in the same way as the honey, because it is hardly split from the wax. In order to be collected, it has to be intercepted before entering the hive. Thus, the apiarists use gadgets that are placed at the entrance, to remove a part of the pollen collected by the bees. These gadgets, called 'catch-pollens', have a simple functioning, based on entrances calibrated for the bees. So, the bees passing through the grid lose part of the collected pollen, which falls in a drawer placed under it. The catch-pollens are placed in Spring, when the swarm of bees is already large. On average, the amount of pollen collected by a catch-pollen varies between 50 and 250g, sometimes reaching 500g per day. The average tends to stabilize at 50g a day.

The pollen collected in the drawers must be treated, in order to prevent fermentation and mould development. Drying is the easiest process. In general, low temperatures must be used (40°C over 24 hours), so that all the mass is homogeneously dehydrated. With this treatment the pollen loses up to 20 per cent of its original weight and stabilizes in values of approximately 2–3 per cent moisture. The masses do not stick together and do not split up. Because of the impurities coming from the hive, pollen must be selected. Even if only small quantities can be detected, these must be removed, with tweezers or a sieve, to separate the largest from the smallest. At an industrial level, this is performed by sieve machines. The pollen commercialized in the form of dehydrated masses is packed in plastic bags or in honey containers and must be kept in fresh, dry and parasite-free places.

Man uses pollen as a regenerating tonic to improve health (Souza et al., 2004). In general, it can be said that the pollen has a synergic action, due to its components, and that it helps the organism to improve all its functions. Also used in cosmetics, the pollen has a regenerating and nutrient action in the skin, and that is why it is used in this industry, mixed in creams and beauty masks. The problem with these products is that their therapeutic applications are still much too empirical, and consequently, not sustained by clinical experimentation.

12.1.3 The royal jelly

Royal jelly is a product of animal origin, since it is segregated by the specialized glands of the worker bees. It was in the 1950s that royal jelly started to raise interest and to capture a significant position in the bee products' market as a food complement and as a raw material for the cosmetics industry. Its use is grounded mainly in the natural food, where the prevention of diseases represents a primary objective.

12.1.4 The propolis

The propolis results from the gathering and elaboration, performed by the bees, of the resins that cover the sprouts and barks of some plant species. This natural product is made of vegetable elements, transformed by the micro-organisms. The transformation of the gathered resins into propolis still remains a mystery. However, the main tree species from which it is extracted are well known.

Propolis is a thick waxy substance of variable colour ranging from yellow, to brown and black. This product's consistency varies according to temperature. At room temperature (20°C) it has the appearance of a hard mass. As the temperature increases from 30°C this paste becomes softer, sticky and viscous. Between 65 and 70°C the propolis melts. When this substance is heated in a water-bath (bain-marie), its mass splits into two different parts: a first one that is waxy, aromatic and soft, which floats, and another that remains at the bottom of the receptable. The propolis is insoluble in water and partially soluble in acetone and alcohol.

The gathering of this product is done by a restricted number of worker bees (carriers) when they are in the last part of their existence, which is from their 18th day of life. The gathering depends on numerous factors connected with the season, climatic conditions, gathering place, bee species and situation of swarms and hives. The bees that gather the propolis do it during the hottest days and hours, i.e. when the temperature is higher than 20°C. There are species that gather a larger amount of propolis than others. This substance is used by the bees to recover the hive walls, reinforce the honeycombs, fix the boxes, seal the holes, fill the cracks, limit the entries, spread on and varnish the hive's and empty cells' internal walls before the queen lays eggs, in order to eliminate any roughness. This product apart from performing an auto-immunity role in the hive, acts as an anti-viral, antibiotic and anti-inflammatory substance, thus preventing the development of germs and favouring the preservation of the honey and pollen. Moreover, as a curiosity, this product has the function of mummifying small insects which manage to enter the hive. These are killed with the bees' poison and covered afterwards by a propolis layer, thus preventing their putrefaction.

There are two different techniques for propolis production: one is based on the natural gathering and the other is based on the artificial gathering, stimulating its production. In the first technique, based on the natural gathering of the product, the collected propolis has a scarce economic value, because extremely small pieces/amounts of this product are obtained. Using the artificial gathering of propolis, the bees are stimulated to gather it and to deposit it on a substratum where the gathering is facilitated. The propolis taken through this process has a higher commercial value.

Once gathered, it is kept in a fresh and dry place, in order to avoid softening, which would result in large quantities of mass, making its manipulation and commercialization difficult. Lyophilization allows propolis to keep for much longer, similarly to any other food, thus preserving all its biologic characteristics.

The properties of propolis are: anti-bacteriological, viral and mycotic; accordingly the propolis is used for medicine, cosmetics and agriculture.

12.1.5 The wax

From the chemical point of view, the waxes are composed mainly of more or less rich and complex mixtures of ester acid hydrocarbons, alcohols and sterols. The waxes may be of a natural or artificial nature, with certain common characteristics. The origin of natural waxes is varied. There are waxes of mineral provenance, as for instance paraffin, present in petroleum, mineral wax, extracted from the subsoil and the mountain wax, present in lignite. The waxes of vegetable origin are produced by the plant's cellular membranes, but it is only commercially viable to extract a limited number of waxes.

The production process of many of the waxes produced by some animal species is still unknown and also not significant/relevant in quantitative terms. In the animal kingdom, the insects are doubtless the major wax producers, above all the *Apidae* (*Apis mellifera; A. florea; A.dorsata* and *A.cerana*).

Beeswax is not a substance found in nature by the bees, but rather a secretion of special glands, located on the underside of the abdomen. It is produced by the worker bees, which are endowed with four pairs of glands formed by a layer of excreting cellules that develop from birth to the 19th day of life. From ancient times, the gathering of honey has taken place along with the gathering of wax, because that was the only material of that particular type available and used in various ways and forms. Nowadays and with the most diverse techniques developed, alternative materials or waxy substances have come on the market at low cost that have widely replaced beeswax in many ways. However, the product's availability has not increased as a result of a technical change of bee products. With the diffusion of intensive bee-keeping and the abandoning of apiculture, wax production has been decreasing strongly. The main users are the apiarists, who use it for the construction of their beehives, according to certain apicultural rules.

Beeswax is largely used in the cosmetics industry, since characteristics attributed to beeswax increase the product's value. The pharmaceutical industry is also a consumer, for the preparation of ointments, pomades and medicinal catheters. In agriculture, the wax is used to improve the fruit's appearance and preservation. Apart from this, the wax is also used in the paper industry, in furniture polish, in candles, in art and in jewellery.

12.1.6 The poison

The bee poison is a mixture of substances produced by the poisonous insect's specialized glandular tissues and introduced in its prey's or enemy's body by means of a stinging apparatus with the aim of paralysing or killing it. To be efficacious the poison must cause pain, damage or any other pharmacological or sensorial activity in the potential predator.

In the case of the social hymenoptera, the poison has a merely defensive function. The bees utilize the poison not to catch their victims, but rather to protect their societies from the predators/ intruders. The interest in the hymenoptera poison is based on two aspects: it concerns, firstly, the clinical problems produced by the poison, particularly the allergic responses; and secondly, the pharmaceutical utilization of the poison as a source of potential useful substances for medicinal, industrial and biological research.

12.2 The honey productive structure

According to FAO data, in 2005, the worldwide production of honey was estimated at about 1 500 000 tons, an amount which is continuing to increase, although with a

Table 12.1 World production of honey per continent (in 1,000 tonnes)

Continent	2001	2002	2003	2004	2005
Africa	145	153	152	152	154
North America	193	188	191	187	188
South America	127	132	137	133	133
Asia	458	497	525	543	545
Europe	311	294	320	328	332

Source: FAO (2005).

Table 12.2 World production of honey per country (quantities and prices)

Rank	Countries	Production in $1000	Production in 1000 tonnes
1	China	522,849	82,000
2	United States of America	142,912	82,000
3	Argentina	139,426	80,000
4	Turkey	128,846	73,929
5	Ukraine	105,445	60,502
6	Mexico	99,007	56,808
7	Russian Federation	92,370	53,000
8	India	90,627	52,000
9	Ethiopia	67,970	39,000
10	Spain	64,485	37,000
FAO estimate			

Source: FAO (2005).

Table 12.3 European production of honey per country (quantities and prices)

Rank	Countries	Production in $1000	Production in 1000 tonnes
1	Spain	64,485	37,000
2	Hungary	35,728	20,500
3	Romania	33,462	19,200
4	Germany	29,628	16,400
FAO estimate			

Source: FAO (2005).

stabilizing tendency (Table 12.1). In the decade of 1979 to 1989, a strong increase of about 30 per cent had been observed, while in the next ten years production augmented slightly, by about 3 per cent.

Per country, China, the United States and Argentina are the top world producers. These countries account for more than 50 per cent of the production and are also the great world exporters of honey (Table 12.2). As can be seen from Table 12.3, the largest honey producer countries within Europe are Spain, Hungary, Romania and Germany.

In Europe, traditionally, a beekeeper considers the need to exploit, on average, at least 150 beehives, in order to obtain the minimal profitable revenue from this activity. However, this figure varies enormously as a function of the environmental conditions to which the bees are subjected: geographical areas that produce intense floral seasons, agricultural biologically balanced surroundings and stable weather conditions are the natural factors that decrease the potential risks of sudden losses, a major reason for unexpected costs. Other factors that increase producer revenues are linked to the possibility of creating scale economies. This may be achieved when exploiting a large number of beehives and if the product has a multifloral standardized quality. In this case attentive control of the huge bee population is essential and technological productive processes are required.

The United States has settled basically in this production model for honey, and, consequently, has a very powerful lobby of honey producers. Their organizational capacity is significant and, just like any modern industry, their level of networking and coordinating joint actions are essential tools for risk control. Also, legislation exists according to their respective requirements.

In Europe, beekeeping is still a less organized professional activity, but much has been achieved in the last two decades. Data of the apicultural census, 1992, indicated that the total number of beekeepers in the European Union (then with 12 Members States) was about 435,000, of which only 13,000 were registered as professional and exclusive producers. In 2000, this number increased to 460,000, of whom 14,520 were professionals. The rise in the professionalization rate of about 10 per cent seems to indicate that beekeepers are becoming more aware of the complexity of their activity and also better prepared to face the many difficulties associated with beekeeping. Since 1997, the European Commission has taken up the challenge of regulating the production processes in conformity with strict food safety requirements.

From the set of joint actions, beekeepers have realized how much they could benefit, in terms of financial and informational support, if they could organize the sector better. Starting from associative groups and then amalgamating into national federations or leagues, the apicultural sector is clearly tending towards restructuring. At the moment, professional beekeepers exploit more than 3 million beehives (excluding the Eastern European countries), representing 40 per cent of the total existing number of beehives in Europe, with Spain, France and Greece having higher proportions of professionalized beekeepers. At present we have no data for the Eastern European countries, which are also very important producers.

Beekeeping is an agricultural activity that is particularly suited to the agroclimatic (edafoclimatic) conditions of developing countries and combines two very useful properties: on the one hand, the activity requires minimal space and can be carried out even by landless people; on the other, it increases rural incomes without large amounts of investments. Therefore, the FAO's Agricultural Services Division has been promoting beekeeping as a compromise to achieve specific development targets: part- or full-time rural employment; a rise in nutritional levels; increases in agricultural production as a result of crop pollination; increase in per region added value (the products of beekeeping, such as honey, beeswax and other by-products are highly appreciated); and conservation of the natural environment.

To stimulate the introduction of more efficient but simple methods of beekeeping in the developing countries, the FAO has expanded its programme to promote apiculture among rural communities in recent years. At present, there are 20 FAO projects operating in developing countries in the regions of Africa, Asia and the Pacific, Latin America and the Caribbean, and the Near East. One-third of these projects are located in Africa. The objective of the majority of these projects is to improve the production of honey through the introduction of more modern techniques to the rural areas. An example of such a project is one in Ghana where the services of apiculture experts and the training, at home and abroad, of extension staff and beekeepers, as well as the supply of equipment and materials, are improving local apiculture and raise the living standards of rural families. This project is also strengthening the training programme of the Apiculture Promotion Unit, attached to Kumasi University, extending its programme over a much wider rural area, thus bringing to many more isolated villages openness, new contacts and technological modernization.

12.3 Consumption and trade of honey in the international market

Based on data from the USDA, the average annual consumption of honey is 220 g per person per year. Some countries surpass this figure significantly. This is the case for Japan (2000 g), Austria (2000 g), Germany (1000 g), the USA (1000 g), and Spain (800 g). The world market of honey is characterized by the presence of two well-differentiated food products: table honey and industrial honey. Therefore, the respective marketing channels are differently designed. Table honey is used for home consumption. In this case, the quality is evaluated based on its origin, colour, texture and botanical variety. Most of the honey consumption is of this type. It has been calculated that 85 per cent of the total distributed amount of honey is used for home consumption, as table honey.

Industrial honey is used in the food industry (bakery, pastry, breakfast cereals and drinks), as a sweetener and perfume and in the pharmaceutical industry or as a cosmetic. We may find it also in tobacco production. In part because of the lowering of prices, but also because of the added value that the mention of the ingredient in the label imparts, honey is being increasingly used in the food industry. Italy is the European country with the largest percentage of used industrial honey: 40 per cent of the total commercialized product.

In the European marketing channels for honey, we find as main segments the producers, the retailers, the importers, and the industrials. Sometimes producers and retailers integrate different functions. This means that it is usual to find producers selling their branded honey, or retailers combining importing activities, for example. In any case, frequently, after being packed the honey reaches the consumer through the normal circuits of the food distribution. However, a large part of the produced table honey (35.7 per cent) is sold directly to consumers. This is a very important figure, as it represents the willingness of European consumers to buy a product which is traditionally connected to local identity expectations, in spite of its availability in the modern distribution stores.

The international trade in honey reflects the reduction in its production by about 25 per cent. China is able to export 37 per cent of its production, mainly to Japan, the United States and Germany. It should be mentioned, however, that the Japanese industry is undergoing a process of substitution of honey by synthetic sweeteners in the composition of certain widely consumed lemonades.

The European Union imports 47 per cent of the honey produced in the whole world. This amount is mainly shared by Germany (with 85 000 tons imports) and the United Kingdom (with 21 000 tons) which together received almost 70 per cent of the total imported amounts in the year 2000. Since the end of the 1970s, the world-wide imports have improved/increased regularly, for a number of reasons: 1) the increases in consumption of natural and dietary products; 2) the dynamism of some operators introducing special varieties of honey; 3) the lowering of prices, due to scale production; and 4) the growth of the industrial use of honey in food processing.

Europe has a heterogeneous productive and commercial structure for honey. The European Union imports about half of the honey that it consumes and is polarized regarding consumption and location centres. Germany, Denmark and England are consumption-oriented countries, whilst Portugal, Spain, Greece, Italy and France are production-oriented ones. The first group imports about 80 per cent of their consumption needs and the second group needs to import just 20 per cent, to satisfy their respective internal market needs.

If we observe the evolution of recent European trade tendencies for the period 1999 to 2003, we can conclude that the total imported amounts of natural honey from the European countries have slightly decreased, although Germany, Italy and Portugal have imported more. This variation was mainly the result of 4 per cent increase in imports from other European countries and of an approximately 2.5 per cent decrease in imports from non-EU countries.

The main countries responsible for the significant amount of imports in Europe have not changed in recent years, and are: Germany, England, France, Italy and Spain. During that period some countries showed an increasing tendency to import larger amounts of honey from Europe, as is the case of the Netherlands, England, Spain and Portugal. Still, this has not been the choice of Greece, Germany, Italy and Denmark, for which an increasing consumption relies on imports from non-European countries.

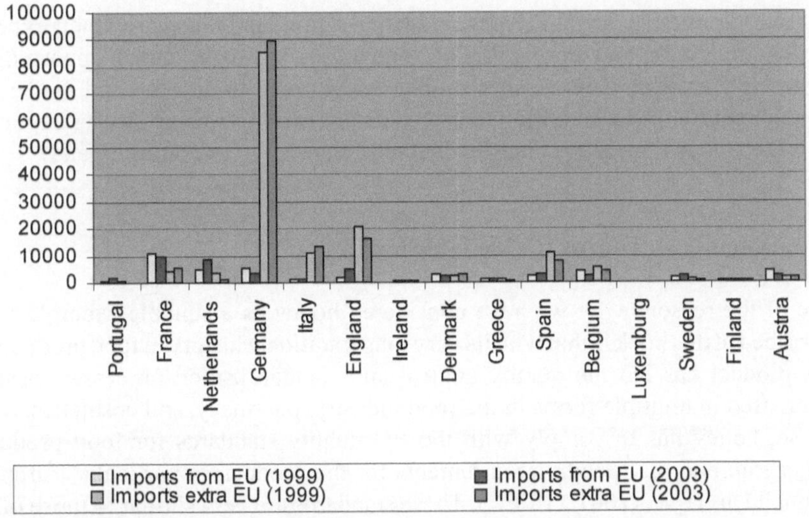

Figure 12.1 European imports of honey

Source: FAO (2005).

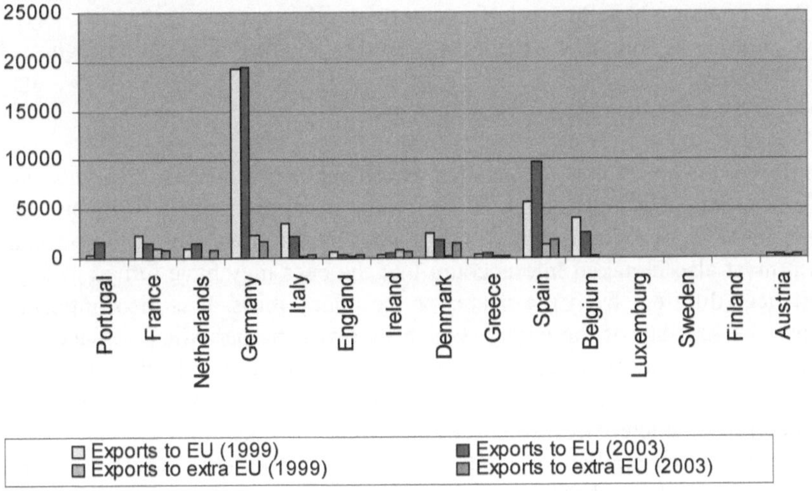

Figure 12.2 European exports of honey

Source: FAO (2005).

All together, Europe imports far more honey then it exports. In 1999 this difference was 145,607 million tons, significantly decreasing to 139,728 tons in 2003. The Iberian countries and the Netherlands were responsible for this positive variation in the trade balance, as they significantly increased their exports, particularly Spain and Portugal. These countries are using many opportunities related to European Commission support programme to develop and modernize their apiculture. It is also important to remember that these countries are exporting mainly to other European countries (namely, Germany, Italy and England).

12.4 Quality standards for tradable honey

One of the reasons why we have considered honey as a suitable example to be described in this book is that with its very long traditional history in food production, this product has led the quality control and standardization processes, besides being used in multiple forms in the food industry, pharmacy, and cosmetics.

So, honey has to comply with the EU quality standards for food products, in general, and the specific requirements for the product, in particular European Council Directive, (2001/110/EC). These standards are very similar to those of the Codex Alimentarius, the Commission's Worldwide Standard for Honey and are requirements demanded by all the major importing countries, namely:

1. It must not have any objectionable flavour, aroma or taint, absorbed from foreign matter during the processing and storage;
2. It must not have begun to ferment or be effervescent;
3. It must be free of any residues caused by medical applications against bee illness;
4. It must not contain any foreign sugar.

In order for the honey to be considered a high quality product, it is worth stressing the precautions that must be taken at the time of its production. They are related to the localization of the apiary – this must be far from any pollution source, and it must also be taken into account that the bees may have turned to sugared substances different from the nectar or the honey mass. It is also important to control the keeping of the apiary, which involves the periodical replacement of queen bees and old honeycombs, as well as the correct use of the fumigator, in order to prevent the excessive use of smoke from jeopardizing the product's organoleptic characteristics, should also be carefully taken into account.

The measures applied to control the bees' pathologies, the nature of the recipient materials, and the hygiene issues associated with the partial taking of honey are particularly important precautions to ensure the product's final quality (Morlot and Beaune, 2004). To render the product fluid, it is necessary to bear in mind that thermal interventions always cause a degradation of the product. This phenomenon ranges from the loss of aromatic substances to the actual damage of the product, which is more evident the higher the temperature and the time taken in applying the thermal treatment. In principle, a temperature of 40° C may be considered in itself not prejudicial to the honey. However, if it is applied over the course of several days, the damage may be higher than that caused by applying temperatures of around 70° C for a few minutes. Another risk that can damage the honey is the excess moisture. In order to maintain the honey in good condition, and therefore to be defined as of good quality, the water content must be lower than 18 per cent.

Just like coffee, honey has also been selected as a product to be commercialized by many fair trade organizations (Fair Trade Standards Honey Small Farmers' Organizations, 2005, in http://www.fairtrade.net/). In such cases there is also an important effort to establish clear criteria to define quality standards. Honey is classified on the basis of its water content and the existence of hydroxymethylfulfural (HMF). In these cases, the basis for the pricing is the commitment of Fair Trade organizations to offer a price covering all production costs, including a comparatively good remuneration of labour (in order to provide adequate living conditions for families, margins for the producer organizations to pay for supporting services to beekeepers, and other social development activities).

As for other traditional products, one of the most used forms to improve acceptance and guarantee the set of quality characteristics of a specific honey is the identification of its origin. Particularly for traditional food products, this link to the locality of production represents an attribute with added value that marketing and labelling tends to increasingly valorize, including in a global context.

In order to facilitate and regulate consumption preferences, the European Commission has accepted the registration of geographical indications and designations of origin, facilitating both the identification of certain products with certain European regions and the process of how agricultural and agro-food products can apply to qualify for such designations. Similarly to wine, some producer organizations are exploring the possibility to develop their products within such a regulatory framework, including for honey.

But a much more serious problem, also related to its origin, is the possibility that honey might be contaminated with genetically modified crops – as things stand at present, bees are not able to distinguish the different biochemical origins of the flowers. The EU Commission has established several guidelines for tacking this problem, common to several other agricultural or animal products. The opinion of beekeepers' associations and traders is that with the implementation of a new form of testing – extraction of pollen and PCR testing of the isolated material – minute traces of GM material are bound to be found, not only in honey, but in all fresh foodstuffs in contact with the wind. However, these horticultural products can already be sold provided the GM levels fall below the legal 1 per cent guideline (the EU plans to reduce it to 0.5 per cent). Since honey is classified as an animal product, in spite of the fact that it originates from vegetation, beekeepers claim that contaminated pollen only occurs accidentally and very occasionally.

Despite this view, honey is still included in the list of products which contain accidental or unavoidable contamination from GM plants (KOM, 2001/425). Thus it is not required to be labelled or licensed as containing GM material – providing it does not reach the maximum 1 per cent limit.

In fact, producers are correct when they say that, considering GM trials are taking place all over the world with many very advanced programmes and considering that pollen can be transported by wind up to 100 miles away, it is only to be expected that in the near future it will be impracticable to organize bees to discriminate between crops.

12.5 Organizing honey production: Three illustrative cases

In recent decades, associative forms of organization have proliferated, integrating many beekeepers, in an attempt not only to organize the productive and marketing function but, mainly, to increase the amount of information related to the scientific and legal requirements that the activity requires in order to produce with as low a risk as possible. Indeed, bee diseases, or plagues in the botanic species, may put at risk years of investment and dedication. Today's biggest struggle faced by most of the beekeepers organizations is linked to the negative effects that pesticides and other chemical products used in agriculture are creating for an huge number of bee swarms, leading to sudden death of the bees or to their inability to reproductive.

In the United Kingdom, the second largest European honey consumer, 25,000 tonnes of honey are consumed each year, but just 1,500 tonnes can be produced by British beekeepers. This means that, in order to satisfy consumer demand, 90 per cent of consumption needs have to be imported. With the EU ban on Chinese honey, which represented 40 per cent of total imports, new sources need to found. This can be a very important challenge for newcomers in the European production of honey: the Iberian countries, Portugal and Spain are already Britain's long time trade partners.

Most of the British honey industry is represented by the British Honey Importers and Packers Association (BHIPA). This association comprises all major importers and packers, including those who buy British honey. Its function is to provide adequate trade and consumer communication across the industry, in close cooperation with BBKA and BFA, including on particular issues affecting the whole industry, such as genetically modified (GM) crops and veterinary drug residues. To achieve this goal, they maintain governmental lobbying and EU representation.

This association has developed extensive programmes related to the control and explanation of the implications of genetic modification, related to both present and future issues. They have developed clear measures and targets to protect bees from contacting with GM crops, such as: instrumental forms in establishing the 'six mile radius' rule in the UK because of retailer demands; taking steps to improve traceability; and monitoring on a regular basis. Also, one of their goals is educate consumers about the process of honey-making, for a better understanding of how honey will always be a 'safe' product.

Less developed countries often rely on apiculture as a source of revenue for local people and self-subsistence. Therefore, in Africa, for example, apiculture should be understood as a productive segment within a local development context. In spite of this and the fact that some national governments have beekeeping institutions and extension services, these are gradually being closed down and eroded through underfunding. Beekeeping is difficult to categorize, unlike mainstream agriculture, and is often overlooked as being a marginal poor-persons activity not worthy of investment and attention.

Very interesting experiences, related to associative forms based on apiculture as a development tool, are the projects 'Linking Local Learners' and 'Bees for Development'.

In the first case, the main goal is to build up knowledge-management strategies for effective rural development in East Africa. Implemented in Kenya, Uganda and Tanzania, this project was supported by the International Fund for Agricultural Development (IFAD) and showed how, based on the local beekeeping, it was possible to explore a knowledge management system that helped local people to find new revenues and use different concepts like: 'know-how', 'peer-exchange' and 'demand driven services'. The action was directed towards small farmers, local service providers and market intermediaries, focusing on the concrete operations of getting small farmers better access to markets, even after programme funding ceases.

IFAD expects that farmers will be able to acquire skills to organize themselves, to share ideas and experiences that will enable them to diminish their poverty. In adjusting their own advances towards knowledge management in a changing global context of modern information and communication technology, farmers need to control the process of knowledge management. Networking should link local service providers, demands on those services, and markets for final production. The project embraces several levels of farmers' organizations: namely, FAO/Farmer Field Schools in all three countries, the farmer, unions UNFFE and KENFAP, as well as the TMLA and numerous independent small farmer organizations

The NGO 'Bees for Development' (BfD) is setting the goal of strengthening livelihoods and fighting poverty in developing countries, and operating support services, particularly organized to provide knowledge and guidance, especially to poor and marginalized beekeepers. The organization, based in England and involved in actions in Uganda, Tanzania, Kenya, Ethiopia, Ghana, the Caribbean Region and Thailand, seems to provide unique insights into the trends and challenges of this neglected sector and can provide the hub of a valuable sharing and learning network for the beekeeping community. There are other NGOs to support beekeeping, but for these NGOs, beekeeping is often a secondary activity, and so they joined 'BfD' for detailed advice and information.

A very different type of organization is CATIM, a technological centre exclusively dedicated to technological innovation, with the target of reorganizing and developing the honey industry in southern Portugal. In its three years of existence, CATIM has reoriented European INTERREG funds to build an extensive set of infrastructures and other equipment, able to promote partnerships for technology and knowledge transfer, specifically related to honey and its by-products. As a technological centre, it is expected to create a pool of excellence able to develop innovative techniques and connect with other technological centres for knowledge creation and diffusion. At the moment, besides its regular interaction with the university, the organization has already established several partnerships with other institutions related to research in apiculture at the international, national, regional and local level.

One of the most important tasks that CATIM is promoting is the on-line diffusion of all the legislation related to honey quality and food safety for producers and traders. Further more, this aspect has been reinforced by several activities to help starter bee keepers and young farmers.

A most interesting example of national efforts in connection with food safety in the sector is the B-Qual Australia Pty Limited, a company founded by the Australian Honey Bee Industry Council. Its function is to independently develop and audit a food safety programme (Honeybee Industry Council of Australia, 2004), in order to accredit Australian's honey quality. In the project of the quality assurance programme, more than 90 per cent of the Australian honeybee industry is represented. Quality standards are to follow both the specific requirements of the European Union (EU) for the export of honey and honey products (including organic) and the Food Standards Australia New Zealand (FSANZ) Food Safety Standard, which requires food businesses to develop a Hazard Analysis and Critical Control Point (HACCP) based on a food safety programme and enabling continued market access both in Australia and overseas. The implementation methods will train industry participants in quality assurance standards, organic standards and biosecurity, as well as providing an ongoing third-party audit system.

12.6 Conclusion

Our chapter has drawn the attention to the singular but complex product that honey is. We have discussed its present conditions within international markets, but we also present two important aspects associated with this product: technical innovation and sustainability. Demanding the most up-to-date techniques in order to allow botanic disease control and being used for revenue creation in many peripheral areas, it may also serve as an environmental monitor. We would tend to agree that honey is a genuine contributor for development as its impact has social, economic and environmental effects. In general, beekeepers do not start their own activity having in mind that beekeeping will be their sole source of revenue. Only after having realized that the activity can bring safe profits and represent a reliable economic source, are the beekeepers able to dedicate professionally and exclusively to the production of bees and honey. Although simple, honey production processes can vary from very primitive techniques to highly innovative methods, and, accordingly, the commercial and distribution channels may present different forms.

In recent decades, associative forms have proliferated, integrating many beekeepers in an attempt not only to organize the productive and marketing function but, mainly, to increase the amount of information related to the scientific and legal requirements that the activity needs in order to produce with as low a risk as possible. Indeed, bee diseases or plagues in the botanic species may put at risk years of investment and dedication. Today's biggest struggle faced by most of the beekeepers'organizations is linked to the negative effects that pesticides and other chemical products used in agriculture are creating for a huge number of bee swarms, leading to the sudden death of the bees or their inability to reproduce.

The possibility to induce in producers the need to innovate in the production and marketing processes using modern techniques can lead to the implementation of associative links in defence of their specific interests and qualities which serve in the long term as forms of local growth and development in many regions around the world.

References

Bryant, V.M. (2001), 'Pollen Contents of Honey', *Canadian Association of Palynologists Newsletter,* vol. 24, n°1, pp. 10–24.

Chattopadhyay, G., Malakar, A., Gupta Bhattacharya, S., Ghosh, A. and Chanda S. (1996), 'Pollen and lipid profile of an index multifloral honey sample from the mangroves of Sundarbans, India', *Annual Agriculture Environmental Medicine*, vol.3, pp.139–144.

European Union (2001), European Union Council Directive 2001/110/EC.

Fair Trade Standards Honey Small Farmers' Organizations (2005).

FAO (2005), News and Notes, Corporate Document Repository in *Apiculture development – The role of FAO.*

Honeybee Industry Council of Australia (2004), Project B-Qual.

Krell, R. (1996), 'Value added products from beekeeping', *FAO Agricultural Services Bulletin No. 124*, Rome, United Nations.

Michigan Technical University (2006), 'Service, Adventure, Knowledge and the Environment', School of Forest Resources and Environmental Science, Master in *Peace Corps and Passion.*

Milner, A. (1996), *An introduction to understanding honeybees, their origins, evolution and diversity,* Bee Improvers and Bee Breeders Association.

Morlot, M. and Beaune, P. (2004), 'Antibiotics contamination, situation in seven countries', Paper presented at the First European Conference of Apidology, Udine.

Oddo, L.P. (2004), 'Unifloral honeys: the work of the International Honey Commission of Apimondia', Paper presented at the First European Conference of Apidology, Udine.

Ruoff, K., Iglesias, T., Luginbühl, W., Bogdanov, S., Bosset, J-O., Kilchenmann, V., and Amadò, R. (2004), 'Potential of Mid-Infrared Spectroscopy for the

authentication of unifloral honey', Paper presented at the First European Conference of Apidology, 'Honey, quality control and residues', Udine.
Titěra, D., Haklová, M. and Kamler, F. (2004), 'Production of honey for pharmaceutical purposes' Paper presented at the First European Conference of Apidology, Udine.
Wanderley, P. A., Wanderley, M. J. A., Boiça, A. L. Jr and Pavan, Á. Jr (2004), 'Effect of four types of honey on longevity and reproduction of *Catolaccus grandis* (Hymenoptera: Pteromalidae)', *Ciencia Rural*, vol.34, no.4, pp. 979–983.
Yuyama, L. K. O., Aguiar, J. P. L., Souza, R. C. da S. and Oliveira, F. P. M. (2004), 'Nutritional value of honey and pollen of stingerless bees of the Amazonian region', *Acta Amazonica*, vol.34, no.2, pp. 333–336.

Websites

http://www.beesfordevelopment.org/info/about/index.shtml
http://www.bibba.co.uk
http://www.fairtrade.net/
http://peacecorps.mtu.edu
http://web.uniud.it/eurbee/Proceedings/quality%20control%20residues.pdf
Paulo Alves Wanderley. 1. Maria José Araújo Wanderley. 2. Arlindo Leal Boiça Júnior. 3. Álvaro Pavan Júnior.

Chapter 13

Market Dynamics and Policy Reforms in the Olive Oil Sector: A European Perspective

Samir Mili

Introduction

The world olive oil market has experienced dramatic changes during the last decade which have resulted, inter alia, in the rapid quantitative and qualitative expansion of world supply and demand, and a significant intensification of horizontal and vertical competition along the product's marketing chain. The European Union (EU) is the major participant in the international olive oil market, accounting for 80 per cent of world production, 70 per cent of world consumption, and 50 per cent of world trade (without including intra-EU trade) (International Olive Oil Council 'IOOC', 2006). Besides the purely economic aspects, olive farming has a high social and environmental content which is always taken very much into consideration when drawing up any regional or territorial development policy.

The available information indicates that the future competitive scenario for EU olive oil will be shaped principally by two types of factors: 1) the market dynamics (supply, demand, trade issues) and business strategies (concentration, internationalization, vertical arrangements, innovation, promotion); and 2) the ongoing agricultural and trade policy reform processes (Common Agricultural Policy 'CAP' reform, World Trade Organization (WTO) Doha Round negotiations, and regional and bilateral trade agreements).

Based on these premises, the purpose of this contribution is to provide an overall evaluation of the market change and policy reform processes in the EU olive oil sector, pointing out the main strengths, weaknesses, opportunities and threats that can be associated with their present and expected evolution. This evaluation is carried out under the general hypothesis that the above-mentioned processes are constrained by a series of macroeconomic, microeconomic and regulatory factors on which there is little empirical evidence or which still are the subject of public debate whose outcomes are uncertain, such as the ongoing trade negotiations. It was also assumed that although some of these factors act on a global level – such as the international regulatory framework – others may vary from country to country and even within countries amongst different chain agents. This means that there is a wide spectrum of possible strategies and courses of action for the future.

13.1 Methodology

This research uses the SWOT (Strengths-Weaknesses-Opportunities-Threats) technique which is a common tool for the evaluation of economic and policy processes. The method has enjoyed consistent popularity during the last three decades, and is a recommended framework to facilitate the understanding of the priorities between alternative actions and to improve strategy. By a systematic evaluation of both external/uncontrollable (opportunities and threats) and internal/controllable (strengths and weaknesses) environments, it provides options and guidance for businesses and policymakers to optimize the fit between internal capabilities and external opportunities in order to achieve superior performance (Novice Vic et al., 2004). The information was gathered from a combination of secondary sources and primary data derived from consultations with a representative set of key informants in olive oil policy and markets, drawn from different professional backgrounds (industry, academia, and public agencies in Spain).

Research has been carried out in two stages. The first stage was to identify the key facts about the EU olive oil market and its macro and micro environments. This has included facts on the olive oil supply, demand, trade and business strategies in the EU and worldwide (internal factors), and on the international agricultural and trade policies related to the sector (external factors). These facts define the current situation of the sector, as well as the main determinants of its performance and possible evolution in the future. The second stage has been to synthesize and to integrate information (primary and secondary), and to evaluate it to determine whether these facts constitute strengths, weaknesses, opportunities and threats for the sector in the EU. The SWOT analysis enables all of the findings to be drawn together, thus reducing the large amount of information to a manageable number of key issues.

It is necessary to clarify that this analysis is performed from the EU perspective, since it is evident that what can be considered strength for EU olive oil can also be considered a weakness in another region or country (e.g. North Africa, Australia, Argentina…). Moreover, it should be pointed out that this procedure of aggregating and synthesizing the information to a manageable number of key issues necessarily implies a simplification of certain aspects, and does not take into consideration all the particularities existing at different stages of the production and marketing chain (farms, milling industries, refining and packing industries, distributors). Here also the strength of one chain agent (e.g. producer) may in some cases be a weakness for another agent (e.g. distributor).

13.2 Results and discussion

The main findings are summarized in a SWOT frame: Strengths, Weaknesses, Opportunities and Threats. Considering that in developed countries business strategies are shaped by an increasingly demand-driven food chain, the analysis of the internal factors begins by exploring demand issues. Subsequently, marketing, trade and supply aspects are examined and, finally, business strategies are investigated. Similarly, the analysis of external factors begins with the international agricultural and trade policies related to the olive oil sector, followed by economic, social and environmental variables that significantly influence the current and future development of the sector in the EU.

13.2.1 Strengths

High product quality and positive image Olive oil is currently considered in most markets a high quality alternative to other edible oils and fats. In the last decade, a big effort was made to improve the quality of olive oils produced within the EU (European Commission, 2000). Product quality has several dimensions, such as health, taste, safety, etc. The traditional Mediterranean diet, whose most characteristic product is olive oil, has contributed to low rates of a number of chronic diseases in the Mediterranean region, and has proved to be a model for healthy nutrition (Wahrburg et al., 2002). As a consequence, it is increasingly appreciated and valued inside and outside the Mediterranean region. Olive oil producers, exporters and institutions communicate the healthy characteristics of the different categories of olive oils to consumers in order to increase their awareness of these attributes.

Numerous consumers appreciate the taste of olive oil and associate it with the Mediterranean diet. As a result, they are willing to pay higher prices for olive oil than for other edible oils and fats. Several EU olive oil enterprises are specialised on the production of high quality extra virgin olive oils, which are sold at high price levels, frequently in gourmet food stores or the speciality departments of large retail channels. The principal EU olive oil brands by consumer awareness and market value ('Bertolli', 'Carbonell', 'Carapelli', 'Hojiblanca', 'Borges'…) belong to the world's leading companies. The main EU producing countries also have a positive country image as olive oil producers.

Product diversity In the EU there are four commercial categories of olive oil (extra virgin olive oil; virgin olive oil; olive oil – a blend of refined and virgin olive oils; and olive-pomace oil issuing from different varieties, origins and production systems. The consumer preference towards differentiated, high-quality food produced in the EU is increasingly a barrier to imports from other countries. A growing segment of consumers prefers quality food with certification of origin (Protected Designation of Origin –'PDO', Protected Geographical Indication – 'PGI', organic). The marketing of specific, exclusive varieties of olive oil is also a way of product differentiation, though more efforts are necessary to explain the differences between different products and descriptive categories.

High participation in the world market The EU accounts for 80 per cent of total world olive oil production, 70 per cent of world consumption, and 50 per cent of world trade (without including intra-EU trade) (IOOC, 2006). The concentration of olive oil production in a few EU regions a priori favours efficiency and economies of scale. On the demand side, olive oil tends to be consumed mainly in the producing areas, a tendency favoured by the preference of many consumers for locally produced olive oil. In addition, Italian and Spanish exports represent 90 per cent of total EU exports. The leading position of the EU in the world olive oil market implies, inter alia, a decisive impact on international olive oil pricing. The most representative marketplaces for this product are located within the EU: Jaén (Spain), Bari (Italy) and Heraklion/Messinia (Greece). 80 per cent of the world olive oil production is marketed by these three markets.

High technological level of the processing industry The positive developments in the EU olive oil sector during the two last decades, favoured by the CAP subsidies paid by means of the Common Market Organization (CMO) for olive oil, made possible a considerable improvement in terms of restructuring and modernization of the milling and processing industry. Increasingly, classic discontinuous plants and continuous plants in 3 phases are been substituted by more efficient continuous plants in 2 phases, which in Spain are already used in most mills (87 per cent in 1999). The main result of the technological modernization process is more efficiency in producing and processing activities, and an improved quality of olive oils.

Benefits for the environment and rural development Apart from the purely economic aspects, olive farming has a heavy social and environmental content. The olive tree is robust, well adapted to difficult climatic conditions in the Mediterranean regions, and makes efficient use of scarce water resources, although it is not very resistant towards frost. Olive oil production is mostly concentrated in the less favoured regions of the EU. With a few exceptions (Tuscany in Italy, Catalonia in Spain) the majority of producing regions come under Objective 1 Community Regional Policy, with a relatively lower level of economic development and higher unemployment rates (European Commission, 2003b).

The maintenance of the traditional olive groves is benign for the environment, since these have low soil erosion and high biodiversity. Indeed, the artificial system of an olive grove is similar to the natural Mediterranean ecosystem. On marginal land, olive growing is often the only alternative to abandonment and desertification. When olive groves sited on marginal land are abandoned, usually they are not replaced with other crops and gradually turn into a kind of scrub. If not maintained, the scrub is a risk for summer fires, one of the chief environmental hazards of the Mediterranean regions (Loumou and Giourga, 2003). However, in the EU, the olive plantations cultivated with techniques with respect the environment are increasing continuously. To some extent, production techniques are communicated to consumers as a means of product differentiation (organic, integrated production). Many persons perceive that the cultivation of olive trees has a positive impact on the environment when appropriate techniques are used, and are willing to reward the use of these techniques.

Olive growing has a positive impact on rural development since during the harvest season it contributes to reduce the high unemployment rates in many EU rural areas (Ministry of Agriculture, Fisheries and Food, 2003). In addition, olive cultivation is increasingly becoming a rural activity combined with tourism. In fact, the olive is a major cultural factor in most Mediterranean regions and has a role that goes beyond mere farming. It is the basis for a whole series of social and cultural events related to gastronomy, tourism and the traditions of local populations (European Commission, 2003a). Thus, in some production areas there are organized tours along olive oil routes, museums…, linking the olive tree with the Mediterranean climate and lifestyle.

Improved professional organization Several institutions deal specifically with the olive oil industry at the national and international level. The professional organization of the sector varies significantly between EU Member States because of differences in production and marketing structures, chain governance modes and legal national frameworks. Some of the main institutions are 'Agecontrol' and 'Agencia para el Aceite de Oliva' which control the EU aid to Italy and Spain, respectively. Interprofessional organizations such as 'Interproliva' in Spain, 'Assitol' in Italy and 'Afidol' in France are involved in promotional activities and the implementation of traceability and quality strategies within the olive oil production chain. Other relevant institutions are cooperatives (in distinct degrees of integration) and their associates at regional and national level, producers' and exporters' organizations, associations of milling and processing industries, and research entities such as the 'Instituto de la Grasa, CSIC' (Seville, Spain). The IOOC acts at multilateral level. In addition, a futures market has been implemented in Jaén (Spain) to reduce price fluctuation risks and to increase market transparency. All these institutions facilitate an improvement of efficiency along the olive oil production and marketing chain.

Strengths in marketing strategies and chain management In the following we will briefly analyse strengths in the marketing strategies and chain management of the EU olive oil industry. It should be mentioned that, in spite of these strengths, in many non-traditional markets there still seems to be rather wide scope for increasing demand by adequate marketing measures. Consumer attitudes and behaviour vary significantly between countries and amongst consumer segments within each country. Olive oil producers should try to adapt the product (type of oil, colour, taste, packaging, labelling), promotion, and other marketing variables to the specific circumstances in each target market. One example for this business policy is illustrated by the results of market research for the design of an *optimal* Cretan olive oil destined specifically to meet consumer needs in the French market (Siskos et al., 2001).

Innovation Most companies consider that product innovations are an opportunity for augmenting profits and market share. For instance, the Spanish group 'Borges' supplies olive oils with various tastes (spicy, etc.). In addition, functional products have been successful in several markets such as the dairy market, and are getting more and more relevant in the oilseed market (see Section 3.4). In the olive oil market, until now only a few functional products have been marketed: for instance, extra virgin olive oil with vitamins A and E from the 'ArteOliva' company, extra virgin olive oil with coenzyme Q10 (antioxidant) from 'Hermejor de la Reina', and olive oil with a tomato derivative (Antelo, 2005). In addition, several enterprises ('ArteOliva', 'SOS') have introduced new packaging (Tetra-Prism). This packaging facilitates the transport and storage in the modern retail chains, and improves the conservation of the olive oil, preserving the properties of the product by keeping it away from oxygen and light.

Promotion In many EU regions, the leading olive oil companies and some public institutions have spent substantial financial resources on brand and generic promotion in order to develop awareness and favourable perception amongst consumers in the expectation of further demand expansion (Mili, 2006). In Spain, in 2004 companies spent €12.5 million on olive oil advertising (Antelo, 2006). There, the 'SOS' group accounted for almost half of this amount. Companies, especially in Italy, that prefer the use of traditional pressing systems for obtaining olive oil to the use of more efficient systems (2-phase centrifugation), communicate this aspect to consumers and link it to local history and culture. Many consumers prefer oils obtained by this procedure (Capogna et al., 2001).

Within the EU, there is potential for the improvement of the coordination of promotion activities. For instance, the introduction of a generic 'EU' designation for external third markets might improve the efficiency of promotion campaigns of EU olive oils in emerging markets. When promoting EU olive oils in non-traditional consumer markets, it is advisable to give consumers simple, clear explanations and suggestions for potential product uses.

Diversification of supply in other food products Several olive oil companies offer a range of other foodstuffs, taking profit from scope economies and synergies in logistics, trade and promotion of different product lines, such as margarine, vinegar, oilseed oils, or table olives.

Raising vertical integration and internationalization Several EU companies are seeking to keep strong positions in the international market through the creation of enterprises in both EU and non-EU producing countries. In this way, the EU enterprises can expand in spite of the demand stagnation in the main EU producing countries, by taking advantage of their comparative advantages both in tangibles (technology, capital availability) and intangibles (brand, reputation, management) in the international market, and by having duty-free access to some markets like the US market. In the US for instance, imported olive oils from some Southern Mediterranean countries do not have to pay any duty, while olive oils imported from the EU have to pay compensatory duties because of production aid received within the EU. Some EU companies import olive oil from third countries under the

'Inward Processing Arrangements' and re-export it (after adaptation to destination market requirements) in order to take advantage of duty-free access.

The Spanish 'SOS' group, which is the leading company in packed olive oil worldwide, plans to increase the number of its own olive oil mills to 15 during the next three years. By this upstream integration it intends that 20 per cent to 25 per cent of its yearly total olive oil supplies could be covered by its own mills, in order to be more independent from origin markets and to stabilize olive oil prices on the final markets. 'SOS' has recently acquired an olive oil milling industry in Morocco, and plans to buy others in Tunisia, Turkey and other Mediterranean countries. Moreover, this company has recently tried to consolidate its leading position worldwide by buying the formerly Italian companies 'Minerva Oli' and 'Carapelli'. With these acquisitions 'SOS' has a share of 20 per cent in the US olive oil market.

Many other EU companies have already established contacts, alliances or subsidiaries in foreign markets. For instance, the Spanish company 'Borges' has subsidiary companies in Jordan, Tunisia, Morocco and Chile. Other important olive oil companies such as 'Coop. Hojiblanca' and 'Acesur' also have alliances, own installations and/or commercial delegations outside the EU (US, Tunisia, Morocco, Syria, Russia, amongst others).

Strategy of distributors' brands Increasingly, there are stable collaborations in the distributor's brand market segment between some EU olive oil producing companies and large food distribution chains. Some examples in Spain are the companies 'Sovena' and 'Olilan' that are specialized in the production of olive oils for distributor's brand. In Spain, the distributor's brands account for almost 50 per cent of the olive oil market, in Portugal 23 per cent and in Italy 11 per cent (Mili and Mahlau, 2005). In the United Kingdom, most consumers buy olive oil by the distributors' brand, as producer brand loyalty is low and many products are perceived as interchangeable (García Martínez et al., 2002).

Creation of own retail stores and focus on the HRC segment Companies like 'SOS' are promoting sales of their oils in their own newly created retail outlets (located in railway stations, airports, high-income districts of cities); other firms are increasingly specializing on the market segment of hotels, restaurants and catering (HRC) (Vargas Sánchez et al., 2005).

13.2.2 Weaknesses

Demand stabilization in major EU producing countries Since the early 1990s, production has increased substantially within the EU, while aggregate demand has increased at a lower rate. Therefore, greater quantities of olive oil have had to be exported. Concretely, between the 2000/01 and 2005/06 crop years, annual production in the EU averaged 2,182,600 t, 39 per cent more than the average for the 1990s (1,569,500 t, crop years 1990/91 to 1999/2000). Meanwhile, the respective figures indicate that consumption increased by 30 per cent from 1,495,600 t to

1,949,000 t (IOOC, 2006). Within the EU, in the last decade the share of the main producing countries (Spain, Italy, Greece and Portugal) in total EU consumption decreased from 93 per cent in 1995/96 to 87 per cent in 2005/06.

Higher price and income elasticity of demand in non-traditional markets In non-traditional consuming countries, many consumers consider olive oil as a non-essential product, for which demand tends to be more price and income elastic than in the traditional consuming countries. In the traditional countries, price increases (up to certain levels) only cause a small, temporary drop in consumption, while in non-traditional markets they may cause a long-run shift in consumption to other types of oils and/or to lower quality and cheaper categories of olive oils. However, as mentioned above, price fluctuations are difficult to avoid, at least at the local level, because of climate variations and alternate bearing. Changes in exchange rates also may accentuate price fluctuations. The current revalorization of the euro regarding the dollar has a negative impact on EU olive oil exports, as the high value of the euro raises the price of the EU olive oils in the United States and many other importing markets.

High production and marketing costs The cost structure of olive groves depends on a multitude of variables, in particular their type (conventional, intensive), the lie of the land (sloping, level ground), productivity, etc. For the EU as a whole the data available does not provide a suitable basis for meaningful comparisons. Generally speaking, the unit costs of production are lower for high-density olive groves than for low-density olive-growing. Moreover, for identical yields, production costs per hectare are higher in the case of olive groves situated on slopes (European Communities, 2003a). This applies in particular to the costs of harvesting, which for most olive growers accounts for 50 per cent or more of the total production costs. The costs of harvesting depend on the costs of labour, the type of grove and its density, the yield per hectare and, above all, the degree of mechanization.

Within the EU, olive oil is cultivated mainly in small and medium-sized farms. Some technologies remain inaccessible for smaller farms and on slopes steeper than 15 degrees. In these farms, the lack of mechanization substantially increases the labour costs. The costs imposed by eco-conditionality also tend to increase production costs in the EU. One way to reduce production costs is to achieve higher productivity and regular yields, e.g. by the application of genetic modifications and improved production techniques. Some studies indicate that there is a considerable margin for cost reduction in organic and conventional olive growing farms by increasing technical efficiency (Tzouvelekas et al., 2001). The available data do not allow a cost comparison of olive production in the EU with its main competitors; however, the high protection level of the EU olive oil industry (see Section 13.4) suggests that in many EU olive growing holdings the production costs are at present higher than in other producing countries.

Furthermore, despite the modernization of the producing and processing industries, the atomization of the holdings, structural duality in the processing industry, high costs of labour, and the relatively low vertical integration in many cases still cause high transaction costs. In several EU regions, there is excess capacity in the olive oil processing industry. While the technological improvements implemented in the producing and processing industries made possible a rise in the average product quality, these changes principally affected processes and products. In entrepreneurial management and organization only few innovations have been introduced, and the EU olive oil sector as a whole still has deficiencies with regard to commercial and logistical organization.

Confusing product denominations Although the wide range of categories of olive oil is, in principle, desirable for consumers (see Section 13.3.1), difficulties emerge for not well-informed consumers to distinguish between different types of olive oil and to recognize quality cues. Many consumers do not know the meaning of designations such as 'extra virgin', 'virgin' or 'olive oil'. In Spain for instance, less than 30 per cent of consumers know at least one PDO or PGI of olive oil. In the Spanish olive oil market PDOs are comparatively a less important demand factor than for other foodstuffs. In Greece too, consumer awareness of PDO/PGI schemes is low (Fotopoulos and Krystallis, 2001). This handicap is even more important in non-traditional markets. For instance, UK consumers are often unaware of the distinction between the diverse olive oils and their corresponding appropriate uses, which deters them from trying and tasting distinct olive oil types. In this market, the olive oil demand would benefit from a dual positioning strategy: different categories of olive oil for different markets (García Martínez et al., 2002).

Structural inelasticity of supply Given the perennial nature of olive cultivation, a major feature of olive holdings is the structural inflexibility of their production response, which restricts their ability to take advantage of market opportunities and to reduce supply when prices are low. The decision to install new olive groves has medium and long-term consequences. Newly planted trees do not attain maturity for a variable period of at least ten years. Once the plantations are providing olives, farmers only have few possibilities to adapt their production to market and price fluctuations. When prices are low, in the short run they may only grow olive trees with low input use, leave the olives on the olive trees without harvesting them, or, in extreme cases, grub up the olive trees.

Wide production and price fluctuations Wide production fluctuations, linked to climate variations and alternate bearing are characteristic of olive groves, whereby huge production tends to be followed by lower production in the following year. The fluctuations in production cause price fluctuations in the olive oil market. Moreover, the marked heterogeneity of production systems implies the need to regulate the coexistence of different types of farming.

Interest conflicts between different chain agents In some EU countries and regions there is a lack of sufficient cooperation between different agents of the production and marketing chain in order to solve common problems. For instance, in the Spanish olive oil market, there is frequent confrontation between the interests of farmers, oil mills and producing cooperatives, on the one side, and packers and exporters, on the other. This situation makes cooperation difficult in questions of common interest such as generic promotion and support of exports, or in handling import regulation when tensions in the market appear (as during the present campaign 2005/2006). In the same cases, promotion of regional olive oils is also in opposition to the development of national or EU-wide promotion strategies.

Heterogeneous regulatory frameworks regarding mixtures of edible oils Within the EU, in most olive oil producer countries the marketing of mixtures of vegetable oils including olive oil is prohibited. But in many other EU countries the marketing of these products is allowed. Some companies have developed mixtures of olive oil with other vegetable oils, on the grounds that these new products may improve the product characteristics, especially the consumer price, in comparison with other types of olive oil and competing oils and fats. For instance, the Spanish company 'Borges' is marketing a mixture of olive and sunflower oils in Belgium under the brand 'Optima', which is distributed by GB (a subsidiary of Carrefour). At present, it is not allowed to market this product in Spain.

Weaknesses in the product strategic marketing The EU olive oil sector has traditionally responded more to supply criteria than to policies to promote demand and to capture new markets (Mili, 1999). Until recently there has been a scarcity of efficient export strategies with medium-and long-term vision. In addition to brand promotion, there is a need for more common coordinated promotion and advertising actions in external markets. Generic promotion has traditionally been undertaken by the IOOC with the financial support of the EU. In recent years there have also been several programmes for the generic promotion of EU olive oils within the EU and in some non-EU emerging consumer markets (European Commission, 2006).

Environmental problems Environmental issues are becoming increasingly relevant for European society. Therefore, pollution and degradation of soil, water and air are also playing an increasing role in agricultural planning. In the olive oil sector, the risk of soil erosion due to inadequate cultivation practices in some regions may be considered as a weakness. There have also been cases where new olive groves were established on land that has a high environmental value due to its ecosystem. In many areas, there may be environmental problems arising from the use of irrigation in regions where supplies are limited. Similarly, environmental problems associated with mills are related mainly to water consumption in regions where this resource is scarce.

Scarcity of reliable, homogeneous statistical information Within the EU, statistical data on olive oil production, marketing and consumption are rather scarce and, sometimes contradictory. The contradictions often have their origins in the use of different methodologies when collecting the data. Regarding production, in spite of the improvement of the databases by the increasing use of the Geographic Information System (GIS), there are still considerable divergences in the estimations by various sources of the number of olive trees, areas under olive cultivation, and yields. With respect to consumption, data obtained from household and HRC panels differ significantly from balance-sheet data and professional associations' statistics (Mili, 1996). Direct sales and consumption on the farm, which still account for significant proportions in all producing countries, also make it difficult to obtain reliable statistical data. The scarcity of reliable statistical information is a relevant weakness since it reduces transparency along the production and marketing chain, increases uncertainty and information asymmetry, and makes planning and an exhaustive analysis of this sector more difficult.

13.2.3 Opportunities

Changes in consumer preferences and increasing demand for olive oil worldwide The rising per capita income and changing consumption patterns towards healthier and safer products, favour the quantitative and qualitative increase of olive oil consumption in and away from home. Within the product categories, these trends especially favour the increase of the high-quality categories of olive oil. World consumption has been progressing fairly steadily, without the fluctuations that are more a production feature. Many consumers consider that olive oil is a healthy, natural and good-tasting product. Nevertheless, olive oil only accounts for approximately 3 per cent of total edible oils consumption at world level. This share is much higher in the EU producer countries (65 per cent in Spain and 69 per cent in Italy) where olive oil traditionally played a major role in oils and fats supply.

However, for instance in the United States, the fourth largest consumer country of olive oil worldwide and with a per capita consumption of 0.7 kg, olive oil only has a share of 2.5 per cent in total vegetable oils consumption (Ismea, 2005). There is a large potential for demand increase worldwide even taking into consideration that olive oil consumer prices are higher than prices of other vegetable oils. In producer countries, the greater increase in consumption will mainly be the result of demand from a segment of population that is concerned about the quality of its diet rather than the relative price level. Similar trends can be observed in non-producer countries, although in this case there is also a large potential for growth in consumption in the young, urban population (Mili and Rodríguez Zúñiga, 2001).

Another salient trend in food consumption is the increase of expenditure on food consumed away from home. Because of the growing importance of eating out in most developed and emerging countries, the use of olive oil in HRC outlets is increasing continuously. In Spain for instance, according to the consumption panel of the Ministry of Agriculture, the share of HRC in total olive oil consumption raised from 12 per cent in 1990 to 18 per cent in 2004, largely at the expense of the at-home consumption share (Mercasa, 2005). In 2005, in Spain there was an overall consumption growth of olive oil by 1.8 per cent in comparison with 2004, which has been the result of an increase in olive oil consumption in HRC by 22.3 per cent and a decrease of consumption in households by 2.9 per cent.

Increasing demand for high quality categories Demand increase is particularly high in the case of high quality types of olive oil (extra virgin, organic, with added vitamins or aromas). High quality consumption is increasing in both traditional and emerging markets. For instance, between 1999 and 2004 the share of virgin and extra virgin olive oil increased from 20 per cent to 27 per cent in Spain and from 37 per cent to 45 per cent in Portugal (Mili and Mahlau, 2005). Some high-income countries, such as the United States and Japan, demand principally virgin and extra virgin olive oils. Thus, the value of the EU exports to these countries tends to increase at higher rates than the exported quantities.

High quality demand is also encouraged by others factors such as the increasing attention to health care and population ageing. In Spain, elderly households (housewives aged than 65 years) on average buy 4 litres olive oil more than housewives aged less than 35 years (Mercasa, 2005). Moreover, in many developed countries, there is a tendency towards a reduction of the size of households. In Spain, on average per capita olive oil consumption in small households is more than double that in large households.

Diversification of food and non-food uses of olive oil There is considerable room for growth and diversification in both food uses of olive oil in preserves, dressings and other preparations and in non-food uses, particularly in cosmetics and the pharmaceutical industries. Along these lines, the use of olive oil by-products in the obtaining of sugar, natural antioxidants, etc. and the generation of energy from the by-products of the milling industries may also be mentioned. However, the quantities of olive oil used for these purposes will presumably be limited compared with the traditional uses.

Increase of exports owing to increasing world agricultural trade liberalization Agricultural trade liberalization made possible improved access by competing countries to international markets, especially non-traditional markets. Trade liberalization is expected to continue affecting the international olive oil market. In the last decade, consumption increased only slightly in the main producer countries of the EU, while imports increased almost exponentially in some non-traditional countries. This evolution is one of the main reasons why EU olive oil companies are increasing their internationalization. With average imports amounting to 202,300 t during the period 1999/2000–2005/06, the United States accounts for 38 per cent of the world total, and is by far the chief destination of world exports (without considering intra-EU trade). With imports of 112,700 t (21 per cent), Italy lies at some distance behind the United States. Next come

Japan (5.7 per cent) and Australia (5.4 per cent) (IOOC, 2006). Some countries like the US, Japan and France are very relevant importers in qualitative terms because they import exclusively or mainly extra virgin olive oils. Since olive oil tends to be mostly consumed in production areas, external trade still concerns an average of less than 20 per cent of world production.

CAP reform In 2003/04 the CAP was reformed (Luxembourg Agreement), introducing changes to the former regulation. The reform aims, amongst other things, to make EU agriculture more competitive and market-oriented. Another objective of this reform is sheltering as large a proportion of direct payments as possible within the subsidies classified as *Green* in WTO terminology. The European Commission expects that CAP reform will stabilize farmers' incomes. For many years, incomes have been varying greatly for olive oil producers because of climate variations and alternate bearing. The introduction of production aids gave a certain stability to incomes in the face of weather variations. Now, the decoupling of subsidy payments from production and the transition to the producer basis allows a stability of payments over time. The European Commission also expects that the reformed CAP will consolidate the positive image of farmers in society with regard to transparency, quality, safety and environmental protection (European Commission, 2003b). In this context, a special Eurobarometer survey (European Commission, 2004) reveals that a majority of EU citizens (60 per cent) support the new directions the CAP has taken recently that put more emphasis on sustainability issues (food quality, safety, environment).

A remarkable aspect of the CAP is the special focus on improving quality. Specifically in the olive oil sector, because the olive oil price is higher than for other common vegetable oils, the consumer expects olive oil to be of impeccable quality. The regional, national and EU institutions are aware of the crucial importance of quality as a key factor for sectoral performance. Some measures of the CAP include the improvement of the intrinsic quality (organoleptic, sensorial parameters), combating fraudulent mixtures, and new rules on labelling in order to provide more precise information to consumers. The EU has also recognized many protected extra virgin olive oils (PDO/PGI). These typical oils, which are mainly famous for their sensorial properties, present a complex and specific qualitative profile including both intrinsic and extrinsic attributes (Caporale et al., 2006).

More efficient production in more liberalized markets It is well known that protectionism tends in the long run to decrease efficiency and competitiveness. In liberalized markets, the vast majority of EU olive oil farms should produce competitive raw materials at lower costs while maintaining high quality and environmental standards. Processors also should produce more efficiently than nowadays. One way of improving efficiency is farm expansion. Bigger units mean better exploitation of technology, capital and labour. Another way to gain efficiency is close and stable collaboration between producers and distributors (traditionally adversaries), by means of increased compatibility of their information systems, and tighter coordination of their logistical and organizational inter-linkages (Holweg et al., 2005). The capacity of national and EU public institutions to facilitate the adaptation of farmers and processing industries to the new competitive setting will influence the rapidity of the necessary changes of the production and marketing processes. Consumers should profit from the aforementioned reduction in costs, as this make possible a decrease in final prices.

Specialization on specific, high-value market segments Another alternative (besides, or combined with, cost reduction) to react to world market liberalization would be the specialization on certain market segments, increasing the quality of products, processes and logistics (Mili and Rodríguez Zúñiga, 2005). Some producers and processors will seek to develop closer links with their customers and aim for regional added value, for instance with local specialities. The development of regional markets is an expression of higher consumer needs for safety and quality that will become even stronger in the course of globalization and open new market opportunities for European olive oil producers. Some olive growing holdings and olive oil mills have been rather successful in developing local markets for virgin and extra virgin olive oils. Others expect to increase their sales by benefiting from tourism and selling these products in speciality stores in the domestic and foreign markets.

Some initiatives linked to olive oil culture and traditions are being developed with this purpose in mind. Up to now, in many Spanish olive oil production areas, the marketing of olive oil with PDO/PGI has been successful, although only few PDO/PGIs are well known at national level. However, there is also evidence of problems with the marketing of regional specialities and organic products due to their high price that are partly a consequence of the high certification costs. One example has been the marketing of olive oil with the PGI 'Terra di Bari' in Italy. Actually there are probably too many quality designations in Spain and Italy, and some of them do not reach a minimum size that is necessary for efficient promotion.

Other ways of specialization are the supply of organic products, top-quality olive oil, or olive oil obtained by means of traditional milling systems, favoured by a stable collaboration between producing and distribution companies (see Section 13.1).

Use of compulsory and voluntary certification schemes The EU 'quality strategy for olive oil' and food policy anticipate incentives for enhancing quality. Quality has to satisfy primary requirements such as health and food safety, and secondary requirements such as convenience, services, durability, etc. The EN ISO 9000, which gives general standards to which a quality system has to conform for accreditation, is of direct importance for the food sector. A regulation to improve food quality and safety is incorporated in EN ISO 9000: the Hazard Analysis of Critical Control Points (HACCP). The Directive 93/43/EEC made the HACCP system mandatory for food companies. The HACCP system has been recommended by the World Health Organization and the *Codex Alimentarius* as one of the best systems in order to guarantee microbial safety and healthy food (Vilar Hernandez et al., 2003).

Article 4.4 of the EU Regulation 1.334/2002 deals with the creation and management of traceability systems for olive oil. Some companies such as 'L. Padillo' and 'Westfalia Separator', have developed and installed a technology that allows the use of HACCP in the olive oil industry (Grupo Padillo, 2005; Vilar et al., 2003). The implementation of an HACCP system in the milling industry not only increases food safety, but also increases the share of virgin olive oil in total olive oil production. In addition, companies can communicate their use of these quality management systems to consumers in order to achieve a differentiation from competitors. However, the industry fragmentation and the lack of sufficient information are obstacles for the rapid implementation of the HACCP schemes.

The increasing use of certification schemes that are imposed on suppliers by the modern retail chains (not by governments), as for instance EUREP/GAP, may be considered an opportunity for the EU leadership in international trade (García Alvarez-Coque, 2005). While food safety and quality standards are seen as crucial for maintaining and improving reputation, additional labour and environmental standards are also gaining ground as strategies to forge customer loyalty and to increase market shares. Food standard schemes are likely to be instrumental in shaping the food system, determining how food is grown, processed and delivered (Fulponi, 2006).

Availability of adequate infrastructure and innovations The EU has some advantages regarding the development of new agricultural technologies (mechanization, biotechnology…) in comparison with most new global players in the world market. The highly developed infrastructure within the EU and the technological level of the whole economy (not only the agricultural sector) are also important factors explaining international competitiveness. In addition, the EU industrial, transport and services sectors are rather efficient in comparison with the situation in most competing countries.

Potentialities opened up by new information technologies The use of new information technologies for national and international commerce is an opportunity for EU olive oil operators. It allows producers to provide consumers in many countries with detailed information about the characteristics of their products: origin, varieties used, organoleptic and sensorial properties, potential uses, and so on. In the EU, almost all the big distribution chains are placing heavy stakes on e-business. These firms take advantage of the Internet as an additional business channel that allows them to expand their business to new buyers, as well as being a tool for customization. There are also firms formed by alliances between on-line communications companies and food enterprises, which combine their complementary skills to sell food through the Net. Business to business (B2B) electronic trade is starting to expand in the olive oil industry. The use of information technologies is more widespread in the EU than in most competing countries, although in the long run competitors may also profit from these technologies.

13.2.4 Threats

Cheaper vegetable oils competition Olive oil has to compete to some extent with sunflower oil, soybean oil, rapeseed oil and other substitute oils and fats. Each oil has its unique characteristics, and none are equally suitable for all purposes. In applications where consumers seek specific characteristics, substitution effects are limited. However, where they can be readily substituted, price can be an important determinant of market share (Owen et al., 1996). Worldwide, soybean oil is the most consumed oil (30 per cent of world consumption of vegetable oils). However, consumption patterns of the different types of vegetable oil vary greatly amongst countries (Mattson et al., 2004). In non-traditional countries where consumers are accustomed to use other vegetable oils, eating habits and traditions may make the introduction of olive oil difficult.

Compared with other edible oils, olive oil is an expensive product. At the retail stage, the price ratio between olive oil and common edible oils is in the order of 4:1 to 5:1. Moreover, olive oil is increasingly competing with new products which have been developed in order to improve the characteristics of substitute vegetable oils (functional food). Many consumers may have a better image of these new products with some added healthy components, than of the traditional oilseed oils. These products are more expensive than traditional oils, but cheaper than olive oil. For instance, in France 'Lessieur' has developed mixture of sunflower oil, rape seed oil, grape seed oil and a fatty acid with high content of oleic oil called 'Oleisol', with addition of vitamins D and E, which is marketed with the brand 'Isio4'. In the view of the French consumers, this product ranks second with regard to the healthy image of edible oils and fats, after olive oil and before sunflower oil and other oils and fats (Parras Rosa et al., 2005). In Spain, 'La Española' ('Acesur') offers 'Naturaceites Soy Plus', which adds vitamins, fatty acids Omega 3, minerals and natural antioxidants to a mixture of sunflower and soybean oils. Some studies even suggest that olive oil is not healthier than rapeseed and other vegetable oils (Boztas, 2001). Olive oil producers and institutions would need to continue enhancing the image of olive oil as a healthy, natural and good-tasting product, in order to maintain the higher consumer valuation of this product compared with these new, potential substitute products.

World supply increases at higher rates than world demand Worldwide, supply is expected to increase at higher rates than potential demand (Mili and Rodríguez Zúñiga, 2001; IOOC, 2001). The rise of the olive oil supply stems, inter alia, from an extension of cultivated areas and technological improvements. It may cause price drops if it is not possible to achieve similar increases in olive oil demand.

Presumably, an increase in world production will be caused mainly by: 1) the expected rapid rise in production in non-EU Mediterranean producing countries (Tunisia, Morocco, Turkey, Syria, Jordan); and 2) the New (World) producing countries: in the coming years, probably the highest rates of growth in production will be seen in some new producing countries, such as Argentina, Chile, Brazil, Mexico, South Africa, China, Australia, New Zealand, and the United States.

These countries have comparative advantages for olive oil production, such as the availability of a cheap labour force (except in Australia, New Zealand and the USA), large extensions suited for olive cultivation, adequate structure of farms and public incentives (low-rate loans, other fiscal advantages). For instance, Argentina is only cultivating some 79,000 ha with olive trees, although according to estimations 80 million ha are suitable for olive groves. In California (USA), the expansion of the olive oil industry is based on the use of new varieties, new plantations with high-density (250–300 plants per acre) or super-high-density (600–900 plants per acre), the use of mechanical harvesters, and excellent quality oils with large and varied labels and styles. In the next five to seven years, the Californian production could easily double when the new plantations come into full production (Vossen and Devarenne, 2005).

Furthermore, the creation of transnational joint ventures and alliances by EU olive oil leading groups (e.g. the 'SOS' group) might in the long run cause a reduction of investments within the EU. These processes may become even more accentuated in future due to bilateral trade agreements and the liberalization of the agricultural world markets. In Southern Mediterranean and New World producing countries, both the public and private sectors are favouring modern, competitive plantations. In the EU, policy makers are paying attention to such developments, as they have to take into consideration the interests of huge areas of marginal land, where olive growing is the only feasible activity.

High dependency on public subsidies The Producer Support Estimate (a measure of support developed by the Organization for Economic Cooperation and Development (OECD) in order to estimate the total monetary transfers from consumers and taxpayers to agricultural producers) shows that these transfers account for approximately 50 per cent of the gross income of EU olive farmers (García Alvarez-Coque, 2001; Drogué, 2005). According to an evaluation study based on Farm Accountancy Data network (FADN) figures, a decrease in income equal to the production aid would imply that 18 per cent of the total area under olive trees would no longer be profitable (European Commission, 2003a). As mentioned above (Section 13.2), the subsidies for olive oil production can at least be partly explained by the location of most olive oil production in less developed EU rural areas.

In the context of the 2003/04 CAP reform, the EU has changed domestic support schemes by introducing decoupled farm subsidy payments. Complementary payments take into account social and environmental considerations. Farmers who get these payments may continue to produce although their costs are higher than the perceived prices. On the other hand, more efficient farms will not get complementary payments. The price drops and the uncertainty with regard to the market regulations in this sector may disencourage investments and the modernization of the more efficient farms.

Increase of imports derived from growing world agricultural trade liberalization International trade agreements, especially bilateral trade agreements with third countries, have facilitated increased access by some competing countries such as Tunisia and Morocco to the EU market. EU olive oil imports from these countries are possible under three regimes: 1) the preferential tariffs and quotas regime (the unique significant quota is that allocated to Tunisia: 57165 tons in bulk in 2005, entering with 0 per cent tariff); 2) the inward processing regime (60 per cent to 80 per cent of the total volume of EU olive oil imports, mainly from Tunisia and to a lesser extent from other suppliers such as Turkey, Morocco and Syria; the quantities imported under this regime are high when EU production is low and are destined principally for refining); and 3) the normal (Most Favoured Nation 'MFN') regime, in which ad-valorem tariff equivalents are quite high (65 per cent to 120 per cent, depending on the product category).

The ongoing WTO negotiations anticipate the elimination of the EU subsidies for agricultural exports by 2013 (for olive oil these subsidies have been eliminated since 1998), and a significant reduction in tariffs and in domestic support linked to agricultural production and prices, in order to reduce distortions on the international market. Some agricultural exporting countries such as Brazil, the United States and India are demanding further liberalization of world agricultural markets. By contrast, other countries such as Japan, the EU as a whole and Switzerland, are less enthusiastic about this perspective, as they fear a negative impact on their agricultural sector.

Trade liberalization also concerns non-tariff trade barriers, which in many countries protect domestic producers and/or consumers. For instance, some developed countries, such as numerous EU countries, Switzerland and Japan, would prefer a more explicit admission of the 'precautionary principle' in WTO agreements. Various countries limit imports of certain products by means of barriers, such as quantitative import restrictions, variable import levies, antidumping duties, restrictive customs procedures and administrative practices, heath and sanitary regulations, safety standards, and labelling and certification requirements.

Risk of lower environmental and social standards within the EU As a rule, nowadays in the EU there are higher environmental and social standards than in other parts of the world. However, market globalization might imply a risk of lower environmental and labour standards within the EU in order to be more competitive in international markets. Some companies may avoid the EU environmental and social standards by operating in non-EU countries. Moreover, as the 2003/04 CAP reform provides decoupled payments to the farmers but does not include similar aid to hired labour, wages in the sector may be affected negatively.

Growing bargaining power of the distribution sector Over the two last decades, the rapid concentration of EU food distribution has led to an oligopolistic market structure and an increase of the bargaining power of the retailers with regard to producers and processors. Nowadays, retailers are the key players in the more demand-driven food system. Payment conditions are usually more favourable for retailers. The market share of distributors' brands is expanding at the expense of producer brands (see Section 13.1). In many modern retail outlets (large self-service stores, hard-discounters, etc.), low prices are an important aspect of their business model, with an extraordinary price pressure on the food industry as a consequence. In addition, the retail chains decide which innovations have to be adopted by farmers and processors in order to take account of the changes required by final markets. Retail outlets also have an influence on the prices paid by consumers for the different categories of olive oil. For instance, in the United Kingdom the supermarket margins average up to 40 per cent for olive oil, compared with 15 per cent with sunflower oil (García Martínez et al., 2002).

Other threats For the olive oil sector there are other threats and vulnerabilities stemming from the socio-economic environment. For instance, in many EU rural areas, young persons prefer to search for employment in the services, construction or industry sectors where usually wages are higher. Sometimes, it is difficult to find a workforce for agricultural activities. In addition, the consequences of the monetary policy (revalorization of the euro with respect to the dollar, the yen and other currencies, the increase of interest rates, changing agro-monetary schemes, etc.) could be a threat for olive oil producers and exporters. A weakness of the world economy (e.g. because of energy scarcity) could have a negative impact on olive oil demand, as consumers might substitute olive oil with cheaper products. Furthermore, the increasing evidence of the prevalence of obesity in many developed countries implies that in this population segment persons have to reduce their calorie intake and consumption of oils and fats.

Last but not less important, as in most other agricultural and industrial activities, the olive oil sector will in the long run be affected by climate change. Temperatures in many regions are rising slowly but constantly as a result of harmful greenhouse gases emissions and other factors. This could have multiple effects on olive growing. As a result of climate change, some areas may become more suitable and others less suitable for olive cultivation. However, the available data still do not provide a reliable basis for the assessment of the consequences of climate change.

13.3 Conclusions

In this study a structured survey of internal and external key factors affecting EU olive oil has been carried out. Table 13.1 summarizes the most relevant strengths, weaknesses, opportunities and threats that have been identified and discussed. The SWOT profile suggests that any strategy should allow the sector to take advantage of its strengths and opportunities, and to minimize the impacts of its weaknesses and external threats.

The principal strengths of EU olive oil are related to the product characteristics, the leading position of the EU in the international market, the benefits for the environment and rural development, and the marketing strategies of the olive oil

Table 13.1 SWOT profile for the EU olive oil sector

Internal analysis	External analysis
Strengths	**Opportunities**
• Product quality and diversity, positive consumers' perception. • Preponderant participation in the world market. • High technological level in the processing industry. • Benefits for environment and rural development. • Strengths in marketing strategies and chain management.	• Changes in consumer preferences, increasing demand for high quality and healthier categories. • World agricultural trade liberalization (more export opportunities). • CAP reform. • Compulsory and voluntary certification schemes. • Availability of suitable infrastructure, R&D, IT.
Weaknesses	**Threats**
• Demand stabilization in major EU producing countries. • Higher price elasticity of demand for olive oil than for oilseeds and in non-traditional markets. • High production and marketing costs. • Short-run inelasticity of supply, large production and price fluctuations. • Scarcity of reliable, standardized statistical data.	• Competition from substitute, cheaper oils. • Higher growth rates for world supply than for potential demand. • High dependency on public subsidies. • World agricultural trade liberalization (more imports). • Growing bargaining power of the distribution sector.

companies. The EU olive oil industry should maintain or even improve product quality, diversity and image. Olive oil operators have to focus on communicating these advantages to consumers, especially in non-traditional markets. By using an adequate marketing mix, competitiveness may be enhanced and the EU's position as a world leader may be strengthened.

The EU olive oil industry also has to adopt strategies in order to face its weaknesses and vulnerabilities. In EU producing countries, where demand has almost stabilized in quantitative terms, operators should make efforts to add value by favouring the consumption of high-quality oils. In addition, companies and public institutions involved in the sector should try to increase demand in non-traditional consumer countries both inside and outside the EU, taking into consideration that the price elasticity is usually higher in new markets. They also have to enhance the communication of the meaning of the different types of olive oil.

Olive oil chain agents should try to solve specific problems related to supply chain management, in particular interest conflicts and marketing costs. Producers and processors also have to deal with some aspects regarding supply, such as the structural inelasticity of supply and the high costs of raw materials. Moreover, the elaboration of more reliable statistical data would contribute to increase transparency along the olive oil marketing chain, as well as the capacity for its investigation and analysis.

The increasing demand for olive oil worldwide, as well as the demand for healthy, high quality products, constitutes a positive context for the development of this sector. The last CMO for olive oil reform has introduced measures oriented to improve product quality, to promote consumption and to strengthen environmental benefits along the olive oil supply chain. The increasing liberalization of international agricultural markets provides considerable opportunities to EU olive oil companies. These may increase their exports and/or create alliances with operators in non-EU producing countries in order to enhance their price-competitiveness and control over international markets. Within more liberalized world markets, companies may try to reach a cost leadership or to specialize in specific market segments (local markets, organic products, PDOs). Technological developments tend to favour production within the EU. Special attention should be given to information technologies that are developing vigorously. Their use has helped to speed up business operations and to improve customer service.

The olive oil sector is also exposed to threats, such as cheaper vegetable oils competition, and an increase of the global supply at higher rates than potential demand. Another threat is derived from the high dependency of the EU olive oil producers on public subsidies. In addition, producers and processors have to deal with the increasing bargaining power of the distribution sector. In the short and medium run they also have to adapt their strategies in order to cope with macroeconomic changes (revalorization of the euro with respect to the dollar, fiscal and agromonetary policies), the social and cultural environment (health care, obesity, population ageing) and, in the longer run, climate change.

Finally, it should be pointed out that this analysis seeks to contribute to the ongoing debate on the implications at the product-level (a differentiated product in this case) of the broad change and reform processes under way in the global agri-food markets, and at the same time stimulate new, more focused research on a subject scantily approached in the specialized literature despite its strong importance for numerous EU regions.

Acknowledgments

The author wishes to acknowledge the support of the Project EU-MED-AGPOL, funded by the European Commission under contract SSPE-CT-2004-502457.

References

Antelo, A.B. (2005), 'Los aceites buscan ser más saludables'. *Alimarket Revista*, vol.183, 04.2005, pp. 115–116.

Antelo, A.B. (2006), 'Aceite de oliva, de nuevo por las nubes', *Alimarket Revista*, vol.191, 01.2006, pp. 107–122.

Boztas, S. (2001), 'Olive oil is no healthier than cooking oil', *The Sunday Times*, www.sunday-times.co.uk/news, 25.02.2001.

Capogna, D., Costagli, G. and Alba, J. (2001), 'El proceso de producción del aceite de oliva virgen. Comparación entre España e Italia', *Agricultura*, vol. 833, pp. 784–788.

Caporale, G., Policastro, S., Carlucci, A. and Monteleone, E. (2006), 'Consumer expectations for sensory properties in virgin olive oils', *Food quality and preference*, vol. 17, pp. 116–125.

Drogué, S. (2005), Description and Quantification of the EU Protection for Imported Olive Oil from Mediterranean Countries. EU-MED AGPOL Project Report D13, INRA-INAPG, France.

European Commission (2000), 'The quality strategies for olive oil', Communication from the Commission to the Council and the European Parliament, COM (2000) 855 final, Brussels.

European Commission (2003a), Working Paper of the Directorate-General for Agriculture: 'The olive oil and table olive sector', Brussels.

European Commission (2003b), 'Accomplishing a sustainable agricultural model for Europe through the reformed CAP – the tobacco, olive oil, cotton and sugar sectors', Communication from the Commission to the Council and the European Parliament, CON (2003) 554 final, Brussels.

European Commission (2004), European Union citizens and agriculture from 1995 to 2003, Special Eurobarometer survey, September 2004, Brussels.

European Commission (2006), 'Promotion of EU agricultural products outside the EU', Press Release, IP/06/364, 21/03/2006. Brussels.

Fotopoulos, C. and Krystallis, A. (2001), 'Are quality labels a real marketing advantage? A conjoint application on Greek PDO protected olive oil', *Journal of International food & Agribusiness Marketing*, vol. 12, n° 1, pp. 1–22.

Fulponi, L. (2006), 'Private voluntary standards in the food system: The perspective of major food retailers in OECD countries', *Food Policy*, vol. 31 n° 1, pp. 1–13.

García Alvarez-Coque, J.M. (2001), 'La PAC, la OCM e l'olivicoltura spagnola', in: A. Cavazzani and G. Sivini (eds), *L'olivicoltura spagnola e italiana in Europa*, Rubbettino, Soveria Mannelli pp. 21–41.

García Alvarez-Coque, J.M. (2005), 'Comercio internacional y globalización. Implicaciones para los productos mediterráneos', in: S. Mili and S. Gatti (coord. eds), *Mercados agroalimentarios y globalización. Perspectivas para las producciones mediterráneas*. Editorial CSIC, Madrid, pp. 9–22.

García Martínez, M., Aragonés, Z. and Poole, N. (2002), 'A repositioning strategy for olive oil in the UK market', *Agribusiness*, vol. 18, n° 2, pp. 163–180.

Grupo L. Padillo (2005), 'El sistema de trazabilidad de L. Padillo para la producción de aceite de oliva en las almazaras', *Mercacei*, vol. 41, pp. 124–125.

Holweg, M., Disney, S., Holmström, J. and Smaros, J. (2005), 'Supply chain collaboration: Making sense of the strategy continuum', *European Management Journal*, vol. 23 n° 2, pp. 170–181.

International Olive Oil Council (2001), 'Projected olive oil production and consumption by 2008', CE/R.54/Doc. nº 3, IOOC, Madrid.

International Olive Oil Council (2006), World olive oil figures, www. internationalolive oil.org, Madrid.

Ismea (Istituto di Servici per il Mercato Agricolo Alimentare) (2005), *Evoluzione del sistema agroalimentare italiano*, Rapporto Annuale Vol. 1, Roma.

Loumou, A. and Giourga, C. (2003), 'Olive groves: The life and identity of the Mediterranean', *Agriculture and Human Values*, vol. 20, nº 1, pp. 87–95.

Mattson, J.W., Sun, C. and Koo, W.W. (2004), 'Analysis of the world oil crops market', Agribusiness & Applied Economics Report nº 529. Centre for Agricultural Policy and Trade Studies, North Dakota State University.

Mercasa (2005), *Alimentación en España 2005. Producción, industria, distribución y consumo*, Madrid.

Mili, S. (1996), *Organización de mercados y estrategias empresariales en el subsector del aceite de oliva*, MAFF, Madrid.

Mili, S. (1999), 'The olive oil sector: International challenges and future scenarios', *Olivae*, vol. 75, pp. 8–16.

Mili, S. (2006), 'Olive Oil Marketing on Non-Traditional Markets: Prospects and Strategies', *New Medit*, vol. 5, nº 1, pp. 27–37.

Mili, S. and Mahlau, M. (2005), 'Characterization of European Olive Oil Production and Markets', EU-MED AGPOL Project Report , CSIC, Madrid.

Mili, S. and Rodríguez Zúñiga, M. (2001), 'Exploring future developments in international olive oil trade and marketing: a Spanish perspective', *Agribusiness*, vol. 17 nº 3, pp. 397–415.

Mili S. and Rodríguez-Zúñiga M. (2005), 'El sector de aceite de oliva español. Transformaciones estructurales recientes y estrategias empresariales', in: S. Mili and S. Gatti (coord. eds), *Mercados agroalimentarios y globalización. Perspectivas para las producciones mediterráneas*. Editorial CSIC, Madrid, pp. 93–107.

Ministry of Agriculture, Fisheries and Food (2003), *El libro blanco de la agricultura y el desarrollo rural*. Madrid.

Novicevic, M.M., Harvey, M., Autry, C.W. and Bond III, E.U. (2004), 'Dual-perspective SWOT: a synthesis of marketing intelligence and planning'. *Marketing Intelligence & Planning*, vol. 22 nº 1, pp. 84–94.

Owen, A.D., Chowdhury, K. and Garrido, J.R.R. (1996), 'A market share model for vegetable and tropical oils', *Applied Economics Letters*, 3, pp. 95–99.

Parras Rosa, M., Senise Barrio, O., Torres Ruiz, F.J., Murgado Armentero, E.M. and Serrano Domínguez, C. (2005), 'El sector del aceite de oliva en Francia', in: S. Mili and S. Gatti (coords.), *Mercados agroalimentarios y globalización. Perspectivas para las producciones mediterráneas*. Editorial CSIC, Madrid, pp. 129–141.

Siskos, Y., Matsatsinis, N.F. and Baourakis, G. (2001), 'Multicriteria analysis in agricultural marketing: The case of the French olive oil market', *European Journal of Operational Research*, vol. 130, nº 2, pp. 315–331.

Tzouvelekas, V., Pantzios, C.J. and Fotopoulos, C. (2001), 'Technical efficiency of alternative farming systems: the case of Greek organic and conventional olive-growing farms', *Food policy*, vol. 26, nº 6, pp. 549–569.

Vargas Sánchez, A., Pelayo Díaz, Y. and Lévy-Mangin, J.-P. (2005), 'Commercial strategies in the olive oil bottling sector in Spain: a strategic group analysis', *Journal of International Food & Agribusiness Marketing*, vol. 17, nº 2, pp. 107–133.

Vilar Hernández, J., Stahnke, W.B. and Nuñez Torres, S. (2003), 'Sistemas de análisis de riesgos y control de puntos críticos. Consideraciones para el sector oleícola', *Oleo Revista*, vol. 95, pp. 70–82.

Vossen, P. and Devarenne, A. (2005), California olive oil industry. University of California Cooperative Extension, Santa Rosa, CA, USA.

Wahrburg, U., Kratz, M. and Cullen, P. (2002), 'Mediterranean diet, olive oil and health', *European Journal of Lipid Science and Technology*, vol. 104, nºs 9/10, pp. 698–705.

Chapter 14
Traditional Belgian Beers in a Global Market Economy

Tessa Avermaete and Gert Vandermosten

Introduction

Traditional products are typically produced by small firms in rural areas, using inputs that originate from the region and relying on regional employment for both the production and the services related to manufacturing (Vaz and Nicolas, 2000; Traill, 1995). Moreover, traditional products are based on know-how that is transferred over generations and embedded in a historical and regional context. Strategies to enhance the competitiveness of traditional food and drinks are therefore of growing concern in the debate on revitalizing rural economies and the maintenance of the cultural heritage in the European Union (Committee of the Regions 1996; Forsman and Paananen, 2004; Ilbery and Kneafsey, 1999; Jordana, 2000; Trail 1995; Tregear et al., 2007; Trichoupoulou et al., 2006).

European, national and regional governments have recognized the importance of traditional products and the related 'traditional know-how and practices' (Kiene, 2006). At the macro-level, quality labels to protect traditional products and processes were launched by public institutions. In addition, there are a growing number of initiatives to promote cooperation between the different stakeholders who invest in the safeguarding of traditional products. At the micro-level, firms invest in the exploration of traditional products as a niche and are thereby supported through public initiatives in mastering traditional products. Yet, the question remains: To what extent can these measures safeguard the production of traditional food from the pressure of large firms and multinationals?

Despite political efforts to implement strategic tools for producers and processors of traditional products, evidence in this domain is in short supply (Tregear et al., 1999). Is the current legislation strong enough to prevent the homogenization of traditional products? Are quality labels a strategic tool in marketing or merely a method to evade the strengthened safety measures that apply within the European Union? Are similar firms with the same 'traditional' label competitors, or can they be considered as colleagues in a common battle against mass production?

The aim of this chapter is to explore the strengths and weaknesses of current legislation on traditional products. For this purpose, focus is put on the case of Geuze and Kriek, traditional beers characterized by a brewing process that is embedded within the Belgian regions of Zennevallei and Pajottenland. The chapter is based on primary and secondary data collected in the framework of the European programme on innovation of the rural economy (*Innoverende Plattelands Economie* – IPE) in Flanders. General figures from the Belgian brewery industry and national beer consumption were used, as well as information from in-depth interviews with experts from public institutes and entrepreneurs from the traditional breweries.

The chapter is structured in five sections. Section 14.1 outlines the policy perspective with respect to traditional food products. Section 14.2 describes the economic situation of the Belgian brewery industry, and Section 14.3 details the general tendencies with respect to beer consumption. An insight into the structure of the brewery industry and consumption patterns is required to understand the paradox of innovation and tradition that is characteristic of the beer sector, which is outlined in Section 14.4. The strengths and weaknesses of common legislation, as well as the evidence of innovation in traditional breweries, are discussed. Finally, Section 14.5 concludes, outlines the future perspectives of traditional beers, and indicates the policy implications of the research.

14.1 A policy perspective on traditional products

The recognition of the cultural identity of regions, the enlargement of the European Union, and the lagging economic performance of rural areas lie at the basis of policy measures to safeguard traditional products and processing methods (Committee of the Regions, 1996). Several initiatives, at all policy levels, have been undertaken for the documentation and preservation of traditional products and the 'traditional knowledge', i.e. the knowledge resulting from intellectual activity in a traditional context (Kiene, 2006). Regulation in this domain aims to prevent unfair competition whereby homogenized products use the name of traditional products, profiting from a name and a reputation to mislead consumers.

At the European level, three quality labels were introduced in 1991 in order to (1) encourage diverse agricultural production; (2) protect product names from use and misuse; and (3) help consumers by giving them information concerning the specific character of the products. It concerns Protected Designations of Origin (PDOs), Protected Geographical Indications (PGIs) and Traditional Speciality Guarantees (TSGs). PDO requires foodstuffs to be produced, processed and prepared in a given geographical area using recognized know-how. PGI includes foodstuffs in which a common geographical link occurs in at least one of the stages of production, processing or preparation. TSGs are granted to foodstuffs that possess a traditional character, either in their composition or means of production. In the case of the TSG, a direct link with regional agriculture is not a necessary condition. It should be noted that the application for registration can only be made by a group of producers or processors.

PDOs and PGIs are mainly acquired by food producers in Southern Europe to protect, for example, wines, cheese and olive products. The success of TSGs is limited. Shortcomings in adequate protection and the lack of a definition for the concept 'traditional' explain this phenomenon. Only recently, the European Community defined 'traditional' as 'proven usage on the Community market for a period at least equal to that generally ascribed to a human generation' (Commission of the European Communities, 2005).

Obtaining a European label might be an opportunity leading to a sustainable profit in the long run. However, the process towards recognition at the European level is not evident for small firms where the entrepreneur is often manager, accountant, processor and marketeer at the same time (Avermaete et al., 2003). Responding to this situation in the field, regional governments have undertaken actions to support entrepreneurs in acquiring a label that protects their products with a regional or traditional character.

In the Dutch-speaking part of Belgium, where the Flemish government has authority in the domain of agriculture and regional economic development, initiatives were undertaken to promote regional products. The VLAM (Flemish centre of agro-and fishery marketing) is thereby responsible for the labelling of food products. As a response to the need of small firms, the VLAM introduced the label of regional products in Flanders STREEKPRODUCT.BE. To qualify for the label, five criteria have been defined: (1) made with regional raw material and/or raw materials that could be considered as regional; (2) generally accepted as regional; (3) prepared according to traditional methods; (4) correspondence between the regional product and region of preparation; and (5) a long-term or historical reputation as regional product. In contrast with the European label, the need to apply as a group of producers or processors does not apply for the regional label. In the short run, the regional label can support the marketing of the traditional product at the regional level. In the long run, the regional label can be regarded as a first step in the process towards European recognition and a valuable tool to protect the traditional product.

It should be clear from the above discussion that different definitions are used to define the term 'traditional'. Traditional can be strongly related to the inputs from agriculture, the materials used in the processing and/or the methods of processing. The strength of a specific label depends on the reputation – within the region, as well as beyond – and the type of product and process and/or regional embeddedness. In practice, understanding tradition thus requires an insight into the historical background of a sector and its affinity with the region and local agriculture.

14.2 Historical background of the Belgian brewery industry

The current economic situation of the Belgian breweries is the result of the country's long history of brewing. In the Middle Ages, all the Belgian monasteries had a brewery. In the 14th and 15th centuries, beer became a popular beverage. Around this time, it was commonly believed that it was better to drink beer than water because epidemics like cholera and the plague could be transmitted by water,

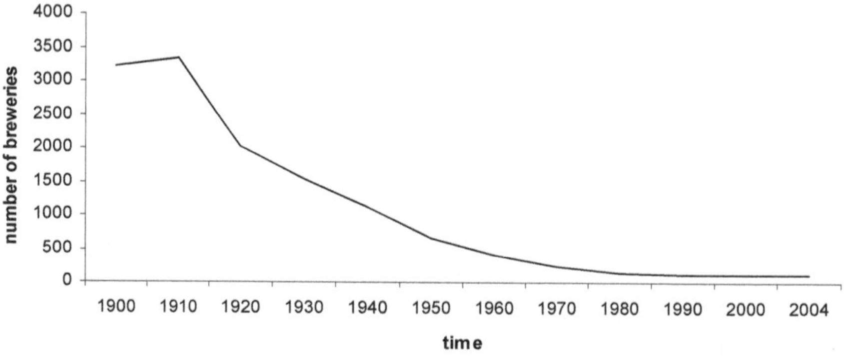

Figure 14.1 Number of breweries in Belgium, 1900–2004
Source: Buelens and Vansevenant (2005).

while the cause of these diseases was eliminated during the brewing process. In the 17th century, almost every town in the country had its own small brewery. After the French Revolution, which caused the destruction of many abbeys and breweries, the brewing industry revived during the Industrial Revolution. The findings of Louis Pasteur (1822–1895) in the domain of yeast and conservation of food were incorporated in the brewing process. On the whole, the quality of the beer improved due to the scientific progress.

At the beginning of the 20th century, Belgium had over 3000 breweries. As can be derived from Figure 14.1, the number of breweries has declined continuously during the 20th century. The World Wars, the lack of raw material and manpower, as well as the economic crises of the 1930s, lie at the basis of the drastic reduction in breweries during the first half of the century. After the Second World War, Belgian had less than 1000 breweries. In the post-War period, many small breweries stopped their activity or became part of larger firms. In 1987, Interbrew was formed from the merger of Brasseries Artois (the second largest brewer in Belgium) and Brasseries Piedboeuf (the largest brewer in Belgium and brewer of Jupiler). Both of the brewers had a history of acquisitions including the Leffe Brand, Dommelsch Brewery and Lamot brewery. Interbrew expanded rapidly taking over many small and large speciality brewers in the country and abroad. In 1991, for example, Interbrew acquired Hoegaarden which is a famous brewery in the country as well as abroad with historical links to regional agriculture. In 2004, Interbrew merged with Companhia de Bebidas das Amérikas (Ambev) to create the world's largest brewery worldwide with a market share of around 14 per cent. As a consequence of the mergers, small breweries were closed and the production of the beers was concentrated in Leuven and Jupille. In this context, the production of Hoegaarden moved away from its historical location keeping only its bottling activities. Nevertheless, the figure below shows that the number of breweries in the brewery sector has been relatively stable over the past two decades.

Table 14.1 Structure and employment of the Belgian brewery industry, 2002

	< 10 empl.		10–49 empl.		50–99 empl.		≥ 100 empl.		Total	
	Empl.	**Firms**	**Empl.**	**Firms**	**Empl.**	**Firms**	**Empl.**	**Firms**	**Empl.**	**Firms**
Number	172	47	1019	47	580	8	4119	13	5890	115
Percentage	2.9	40.9	17.3	40.9	9.8	7.0	69.9	11.3	100.0	100.0

Source: IPV – Detailanalyse (www).

Table 14.1 shows the structure and employment within the Belgian brewery industry. In line with the general tendency in the food industry as a whole, the beer sector is dominated by small and very small firms. Over 80 per cent of the breweries have less than 50 employees. The very small breweries are typically family-based. It should be noted that 47 breweries have less than 10 employees with an average of 3. Despite their small size, some of these firms have a well-known reputation.

There is no doubt that most of the employment in the sector is provided by large firms. Less than 12 per cent of the firms represent almost 70 per cent of the employment in the sector. It should be marked that Inbev is by far the most important source of employment in the Belgian brewery industry. In 2002, the former Interbrew counted over 3000 employees in Belgium. Despite recent reforms in management and dismissals, the company remains the largest brewer accounting for about half of the total employment in the Belgian brewery sector.

Apart from the direct employment in the industrial activity, breweries have close links to other sectors. This indirect employment is much greater than the direct employment and includes jobs in delivery companies, hotels and restaurants, the beer trade, large-scale retailing, transport, etc. In order to guarantee their market share, large breweries have taken over pubs, restaurants and wholesalers.

14.3 Trends in beer consumption

Beer consumption is embedded in the cultural and traditional Belgian heritage. Although Belgium is one of the countries with the highest beer consumption per capita in the EU, beer consumption has decreased over the past three decades. As can be derived from Figure 14.2, the fall in beer consumption has been compensated by an increase of waters and lemonades and a doubling of the wine consumption as compared with 1965. It is evident that the changing drink patterns in the entire Western society result from increased awareness of the ethical, health and ecological aspects of food consumption, the fusion of cultures, and imports of ethical food and drinks. Concerning the latter, it is a fact that the Belgian consumer is inspired by the many nationalities living in the country (Askegaard and Madsen, 1998).

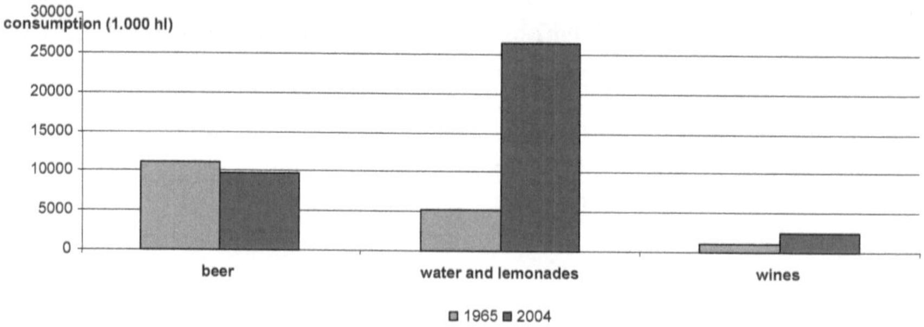

Figure 14.2 Evolution of beverage consumption in Belgium

Table 14.2 Consumption of beer in Belgium, 1999–2003 (in 1,000 hectolitres)

	1999	2000	2001	2002	2003	2004	2005	Change 1999–2005
Thirst quenching	1.233	1.123	1.015	945	881	698	623	- 49.4 %
Pilsner	6.974	7.006	7.027	7.014	7.044	6.631	6.591	- 5.5 %
Other Pilsner	598	533	503	478	463	510	487	- 18.6 %
Abbey/Trappist	739	755	794	816	853	865	891	+ 20.5 %
Degustation	363	365	363	374	452	558	546	+ 50.3 %
Fruit beers	222	214	215	208	242	310	336	+ 51.5 %
Total	10.206	10.064	9.986	9.901	9.935	9.562	9.475	- 7.2 %

Other Pilsner: table beer, non/low alcoholic beers, and luxe Pilsner.

Thirst-quenching beers: Red-Brown, Amber and white beers.

Degustation beers: strong blond, regional beers, pale-ale, stout, scotch-X-mas.

Fruit beers: geuze, kriek, fruit beers.

Source: Belgian Brewers (www).

Table 14.2 shows recent figures on beer consumption in Belgium, with a breakdown for the different types of beers. It can be observed that the popularity of some beers has decreased dramatically, whereas the consumption of other beers has increased significantly. Thirst-quenching beers such as amber, luxe lager (including Jupiler and Heineken), alcohol-free beers, table beer and white beer have lost their market share in Belgium. The consumption of speciality beers, like abbey beers and strong blond beers, as well as the consumption of *sweet* fruit beers, has gained market share.

The observed consumption tendency is a reflection of diverse phenomena, not only in Belgium but also in the rest of Western Europe. The increase and decrease of certain types of beers are in line with consumers' concern about food health, the supply of other alcoholic and non-alcoholic drink industries, the role of food patterns in social status, and ethical considerations.

Scientific evidence and public campaigns on the relation between alcohol and its influence on overall health may have had an influence on beer consumption. Leaving the discussion on the nutritional values of sweet drinks somewhere in the middle, health concerns can be an argument for consumers to decrease beer consumption in favour of waters and lemonades. One can expect that this argument holds particularly true for thirst-quenching beers. The move away from, for example, lagers and table beers is being encouraged by the rise of taxes on alcoholic drinks.

Nevertheless, alcohol avoidance as such does not explain the full picture, as we have seen that wine consumption has been steadily increasing. Apparently, beers have lost market share compared with wines (Figure 14.2). Part of the explanation lies is the availability of cheap wines from all over the world. In addition, the social status related to wine drinking makes people replace beer in favour of wine. Whereas wine has entered the households, strong drinks such as breezers have found their way to pubs and clubs.

The question then arises: How, in view of this pressure on beer consumption, do some beers gain market share. Part of the argument lays in the increased concerns with respect to food safety and recent food crises that have struck the food sector in the European Union. Consumers increasingly want to know where products come from and how they are produced. In this context, specialty beers respond to the consumer's need for food traceability (Ilbery and Kneafsey, 2000; Grunert and Ottowitz, 1997; Tregear et al., 1999). Specific brands on beer bottles, showing the place or a part of the brewing process, enforce this perception of food safety. It should thereby be mentioned that scientific reports still show higher safety risks in the traditional products as compared with the homogenized foods as produced by the multinationals (Jordana, 2000).

However, the core explanation for the increased consumption of specialty beers can be found in the current desire of consumers to differentiate through qualitative aspects (Jordana, 2000). As a reaction against homogenization and mass production, consumers search for products that are original or express a certain degree of authenticity. It is therefore observed that, at this time of globalization and market liberalization, there is an increased interest in local food traditions throughout the European regions (Askegaard and Madsen, 1998; Treagear et al. 1999). Through the maintenance of traditional brewing processes, small-scale production and the historical background, specialty beers fulfil the current desire for authenticity and nostalgia. This is in line with results reported by Caporale and Monteleone (2004). Investigating the influence of information about the manufacturing process, the authors found that information about the product's history and manufacturing process has a significant impact on beer acceptability.

The argument of intrinsic values related to the cultural and historical background holds particularly true for the fruit beers in Belgium as this category includes Lambic and Lambic-derived beers. These beers (Lambic, Geuze, Kriek, Vieille

Kriek and Vieille Geuze) are, in the first place, embedded in a typical geographical area and produced according to a long tradition. Nevertheless, brewers of fruit beers are tending to move away from the traditional brewing process, replacing traditional processes by standardized methods that are less labour intensive. In addition, some brewers of the traditional fruit beers are responding to the demand for sweet foods and drinks by adding sugar. Against this tendency, some Belgian brewers recently combined their expertise and are working out strategies to maintain the traditional products.

14.4 Strategies to remain traditional: The paradox of brewing Geuze and Kriek

14.4.1 Pressure on tradition

The history of Lambic and Lambic-derived beers such as Geuze and Kriek goes back to the Middle Ages. There are three key differences between the brewing process of a conventional lager beer and Lambic. First, to brew a spontaneous fermented Lambic, no yeast is artificially added to the wort, but the wort is exposed to the open air of the Zenne Valley. The result of this method is that wild yeast cells, among them Bretanomyces bruxellensis and B. lambicus, which are always in the open air of the environment of Brussels, come into the wort and start the fermentation in a natural spontaneous way. Because of this spontaneous fermentation, Lambic is not only a regional but also a seasonal beer, which can only be brewed in the winter season (October – May). In summertime, there are too much undesirable bacteria, which can infect the wort and influence the fermentation negatively. Secondly, more than 30 per cent of unmalted wheat is utilized. Third, the Lambic brewer uses old dry hop, aged over three years. He does not want the bitterness of the fresh hop, but the conservation property of the dried hop. The lager brewer uses young hop which gives beer its typical taste (Toelen, 2006; van den Steen, 2006).

In contrast with other beers, Lambic has no foamy head. In order to respond to the consumer's preference, old Lambic was blended with young Lambic. In this way, the production of carbon dioxide is stimulated. The mix of young and old Lambic is stored in Champagne bottles allowing a secondary fermentation.

Because of changes in the rural economy at the end of 19th century, the cultivation of fruit became very popular in the surrounding regions of Brussels. Cherries were added to the Lambic beer, causing a secondary fermentation and the production of Kriek. Currently, the same process is applied using other fruits such as raspberries.

Despite the worldwide explosion of interest in Lambic beers, there is pressure on the authenticity of the original product. Foremost, there is a tendency to organize the brewing process similar to conventional beers. Yeast is added. The original wooden casks are being replaced by stainless steel casks. Spontaneous secondary fermentation is abandoned, and pasteurization is applied. Carbon dioxide is added and the beers can be bottled like industrial products. The result of the changes in brewing is the production of a homogenized beer with a mass appeal, though still carrying the name of a traditional product. Another pressure on the taste of Lambic beers is related to the consumers' demand for sweeter foods and drinks. As a response to this change in the consumption pattern, some brewers add sugar to their Lambic beers. It leaves no doubt that the sweetening of beers is a threat to traditional Geuze and Kriek. Nevertheless, it is a fact that sweetening also explains the drastic increase in the consumption of fruit beers (see also Table 14.2).

Aside from the process and the flavour of the beer, the regional link that is intrinsic to the original product is also under pressure. Brewers outside the Zenne valley and Pajottenland make Lambic beers according to the traditional procedure and apply the names Lambic, Geuze and Kriek. But, as original Lambic beers require specific characteristics of the open air, beers produced outside the Zenne valley and Pajottenland do not fulfil the necessary conditions of the original product.

14.4.2 Labelling as a strategy

Already in the first half of the 20th century, Lambic brewers experienced the pressure on, and the misuse of, the name of their traditional product. The first attempt to protect the product dates from 1930. However, only in 1965, did the Belgian government impose a restriction on the use of the names Lambic, Geuze and Kriek. Spontaneous fermentation as well as the use of specific raw materials became obligatory. Later, the presence of certain bacteria was added, and, in 1994, brewers adding sugar were required to note this on the bottle (van den Steen, 2006).

When in 1991, the European Commission introduced certifications of traditional products, Lambic brewers considered this mechanism as a means to protect the traditional component of the product. In 1997, five groups of beers related to Lambic acquired the label of Traditional Speciality Guaranteed. The labels can now be used by all brewers who fulfil the conditions as described in Table 14.3.

According to the definitions, 'Vieille' is the best guarantee for authenticity as secondary fermentation is a condition to apply the prescript. The use of sugar to sweeten the original product is allowed for all types of Lambic beers with a TSG label. Further, the TSG label does not guarantee that the beers are produced in the original regions of Zenne Valley and Pajottenland. As a result of the broad definitions, the traditional value of beers with a TSG label is very diverse.

Since 2004, brewers of Lambic beers can apply for the regional label STREEKPRODUCT.BE. This label can only be used if the inoculation with micro-organisms takes place in the Zenne Valley or Pajottenland. In contrast with the original goal of the regional government, i.e. to support producers in obtaining the European label, the regional label bestows an added value to TSG.

Table 14.3 Name and definition of Lambic and its derivates that have acquired the TSG label

Name	Definition
Faro	Certain components of Faro have been matured for over three years and give it that characteristic flavour. Faro is relatively acid, and is sweetened with candy sugar
Lambic, Gueze Lambic, Gueze/Lambiek, Geuze-Lambiek, Geuze	The name Lambic is taken from the *B. lambicus*, a crucial micro-organism which gives lambic its characteristic flavour. The oldest component of gueuze or gueuze lambic has matured for over three years in oak casks.
Kriek, Kriek-Lambic, Framboise-Lambic, Fruit-Lambic/Kriek, Kriekenlambiek, Frambozenlambiek, Vruchtenlambiek	A sharp tasting beer, Kriek is the result of blending various lambics. It contains between 10% and 25 % of cherries (as fruit, juice or concentrated juice), like other fruit beers, except Peach which can contain up to 30 %
Vieille Gueuze, Vieille Gueuze-Lambic, Vieux Lambic/ Oude Geuze, Oude Geuze-Lambiek, Oude Lambiek	Gueuze and Vieille Gueuze are the result of a blending of lambics, with the oldest component having been matured for at least three years in wooden casks. One difference is that, in Vieille Gueuze, the weighted average age of lambics used must be greater than or equal to one year. Another distinction between a Gueuze and a Vieille Gueuze lies in the fact that a Vieille Gueuze has undergone a secondary fermentation and is bottled
Vieille Kriek, Vieille Kriek-Lambic, Vieille Framboise-Lambic, Vieux fruit-Lambic/ Oude Kriek, Oude Kriekenlambiek, Oude Kriekenlambiek, Oude Frambozenlambiek, Oude Fruit-lambiek	The distinction between a Kriek and a Vieille Kriek or between a Fruit Lambic and a Vieux Fruit Lambic lies in, amongst other things, the fact that a Vieille Kriek/ Vieux Fruit Lambic has undergone a secondary fermentation and is bottled in lees.

Source: http://ec.europa.eu/agriculture/foodqual.

In total, 13 Belgian brewers produce Lambic or Lambic-derived beers. It concerns mainly very small firms, including eight breweries with less than ten employees. Three larger breweries, with more than 50 employees, produce Lambic-derived beers. Among the larger firms are the only two breweries from outside the original region. In addition, the strategic behaviour of the entrepreneurs is diverse. A number of Lambic derived beers are considered as 'commercial' by their competitors. Consumers' preferences, such as sweetness of the beer, are key forces for innovation to these larger brewers, and market innovation is considered very important. Some entrepreneurs are small-scale traditional brewers who only apply organizational and market innovations as a means to survive on the market. Product and process innovations are based on trial and error, experience and taste preferences. One can think, for example, of the purchase of 'new' casks from wine producers in the South. A limited number of brewers are following up recent findings in brewery science as a means to safeguard the traditional product. They implement product and process innovation with respect to the original character of the final product. These breweries demonstrate very explicitly that tradition and innovation are not a paradox but can, in contrast, bestow an added value to maintain traditional products.

14.4.3 Strengths and weaknesses of the labels

The European and regional labels have, to some extent, reached the goal to protect a traditional product from competition with homogenized products with a mass appeal. The period of maturation and the use of oak casks are made a necessary condition to apply the name 'Lambic' to Lambic or any of the Lambic-derived beers. In the case of Vieille Gueuze and Vieille Kriek, the use of the label requires a secondary fermentation. Additionally, STREEKPRODUCT.BE is an appreciation of the regional embeddedness of the beer.

Apart from the certain degree of protection that is created by means of the labelling, an important side effect of this mechanism was the cooperation between brewers. The European recognition in 1997 was the driving force to establish the Hoge Raad voor Ambachtelijke Lambikbieren (High Council for Traditional Lambic Beers), called HORAL. HORAL has three goals. The organization aims to promote traditional Lambic beers and Lambic derived beers, whereby attention is paid to all the aspects from brewing to the art of serving a beer. Further, HORAL takes responsibility for irregularities and takes measures to protect traditional Lambic beers. The organization therefore responds to a crucial factor in entrepreneurship amongst small firms referred to as 'learning by interaction' (Noeme and Nicolas, 2004). As quoted by a traditional brewer of Vieille Geuze: '*One of the main profits from the European recognition is the co-operation among Lambic brewers*'. The cooperation is extremely important for these small firms where marketing and promotion activities are organized by the manager-brewer. Through the organization of events, such as the Toer de Geuze, the Lambic brewers act as colleagues in a common battle against the loss of the traditional beer.

More recently, the TSG label has proved its role in safeguarding the identity of specific characteristics in brewing that are not in line with current measures on hygiene and food safety. Like cheese and wine production in Southern Europe, the production of Lambic beers requires the use of materials and processes that are in conflict with current legislation. Hazard Analysis of Critical Control Points (HACCP) defines that: 'Equipment should be designed so that all food contact surfaces are smooth, non corrosive, non toxic and free from pitting, cracks and crevices. Floors, walls and ceilings have to be of material that is durable, smooth, cleanable and suitable for the production conducted in the area.' Whereas many commercial brewers have already replaced their wooden casks, brewing in wooden casks is necessary to produce traditional Lambic. Also the porosity of the walls provides a necessary condition for the spontaneous fermentation. In 2003 this situation was the key issue of a debate between brewers and the Federal Agency for Food Safety. Referring to the TSG label and similar incidences with wine and cheese producers in Southern Europe, the Lambic brewers were granted exemptions from the standard safety and hygiene measures.

From a commercial point of view, the label provides a signal to the consumer. However, Lambic brewers argue that profits from the visual appearance of the label are limited. In contrast with products from Southern Europe, such as cheese and wines, the Belgian consumer does not yet appreciate the European labels as an added value of national products. Similar findings were reported by Tregear et al. (1999), who concluded that the awareness of EU certificates in the UK is very low and understanding of the criteria non-existent. Despite internationalization of food markets and the disappearance of national borders within the European Union, it is clear that regional patterns of food-related preferences and behaviour are still apparent (Askegaard and Madsen, 1998).

14.5 Concluding remarks

This chapter has explored the position of traditional beers in a global market economy. On the basis of the figures from the Belgian beer market and in-depth interviews with experts and entrepreneurs, the research demonstrates the paradox of innovation and traditional brewing. The main arguments of the chapter are summarized in Figure 14.3 which shows that current pressure on tradition combined with support measures leads to innovation in the broad sense of the word (Lundvall, 1995).

Three sources of pressure were identified: producers, consumers and government. The remarks in italics refer to relevant examples for the case of Lambic and Lambic derived beers. In the long run, these pressures on tradition lead to product innovation, process innovation, organizational innovation and market innovation. The italicized remarks refer to examples of current innovations in

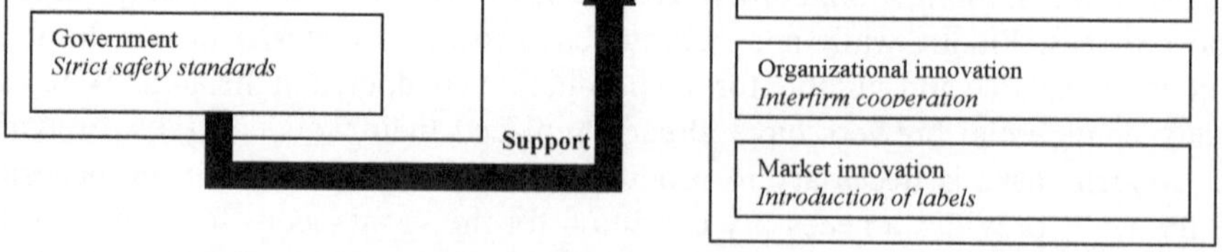

Figure 14.3 Innovation and traditional brewing

traditional breweries. The study demonstrated that governmental support through the introduction of labels provides a driving force, mainly for market innovation amongst small traditional brewers. Nevertheless, the current labelling system has considerable shortcomings and policy makers face many challenges with respect to professionalizing the system.

On the one hand, it seems that labels are often very weak, and therefore even commercial products can make use of the European TSG label. This situation is the consequence of larger breweries lobbying in the process of European recognition. The use of labels for which only a very few constraints are defined may in the long run weaken the entire value of the label. Policy makers should be aware of the consequences of broad definitions and follow up the use of the labels by manufacturers. An evaluation of the size and the market orientation of these manufacturers may already give an indication of the success of the label with respect to its original goals.

On the other hand, labels are only relevant if consumers recognize the label and associate the label with an added value. Studies in the domain of consumers' awareness of labels concerning authenticity are, however, limited. Research of the reputation of products, within the region of origin as well as abroad, can be a useful indication on the economic power of European and regional labels. It can be expected that much more effort should be made to inform consumers about the added value of these labels.

Acknowledgement

This research was carried out in the framework of European project on innovating rural economy (*Innoverende Plattelandseconomie* – IPE). The project is being carried out by a consortium of 18 public and private organizations and aims to professionalize rural suppliers and promote rural activities.

References

Askegaard, S. and Madsen, T.K. (1998), 'The local and the global: exploring traits of homogeneity and heterogeneity in European food cultures', *International Business Review,* vol.7, pp.549–568.

Avermaete, T., Viaene, J., Morgan, E.J. and Crawford, N. (2003), 'Determinants of innovation in small food firms', *European Journal of Innovation Management,* vol. 6, n°1, pp.8–17.

Buelens, M. and Vansevenant, A. (2005), 'Vijf voor Twaalf', *Het Brouwersblad,* 6–17 juni.

Caporale, G. and Monteleone, E. (2004), 'Influence of information about manufacturing process on beer acceptibility', *Food Quality and Preferences,* vol.15, pp.271–278.

Commission of the European Communities (2005), Proposal for a Council Regulation on agricultural products and foodstuffs as traditional specialities guaranteed, 2005/0270(CNS).

Committee of the Regions (1996), *Promoting and Protecting Local Products. A Trumpcard for the Regions*, Brussels, Committee of the Regions.

Forsman, S. and Paananen, J. (2004), 'Value creation in local food supply chains: market opportunities and challenges', in: 2004 IAMA World Food & Agribusiness Symposium Papers and Presentations, 12–15 June 2004, Montreux, Switzerland, 20 p.

Grunert, K.G. and Ottowitz, T. (1997), 'Neumarkter Lammsbräu: brewing beer for Greens', in: B. Traill and K.G. Grunert (eds), *Product and process innovation in the food industry*, London, Blackie Academic & Professional, pp. 99–111.

Ilbery, B. and Kneafsey, M. (1999), 'Niche markets and regional specialty food products in Europe: towards a research agenda', *Environment and Planning A*, vol. 31. pp. 2207–2222.

Ilbery, B. and Kneafsey, M. (2000), Producer constructions of quality in regional speciality food production: a case study from south west England, *Journal of Rural Studies,* vol.16, pp. 217–230.

Jordana, J. (2000), 'Traditional foods: challenges facing the European food industry' *Food Research International,* vol. 33, pp. 147–152.

Kiene, T. (2006), 'Traditional knowledge in the European Context', *Ressources Naturelles,* 02/2006.

Lundvall, B.A, (1995). *National Systems of Innovation: Towards a Theory of Innovation and Interactive Learning*, London, Biddles Ltd.

Noeme, C. and Nicolas, F.M. (2004), 'The process of technological learning and co-operation in small firms', in: M.T. Noronha Vaz, J. Viaene and M. Wigier (eds.), *Innovation in Small Firms and Dynamics of Local Development*, Scholar Publishing House, Warsaw, pp. 122–162.

Streekproducten Vlaams Brabant (2005), Jaarverslag.

Toelen, T. (2006), De Graanfluisteraar: Boeren, Bakkers en Brouwers in de Westhoek. Centrum voor Agrarische Geschiedenis, pp. 107–135.

Trail, B. (2005), 'Small and medium sized food manufacturing enterprises in the EU',. Working Paper Reading, University of Reading.

Tregear, A., Arfini, F., Belletti, G. and Marescotti, A. (2007), 'Regional foods and rural development: The role of product qualification', *Journal of Rural Studies*, vol.23, pp. 12–22.

Tregear, A., Kuznesof, S. and Moxey, A. (1999), 'Policy initiatives for regional foods: some insights from consumer research, *Food Policy*, vol. 23, n°5, pp. 383–394.

Trichopoulou, A. Vasilopoulou, E., Georga, K., Soukara, S. and Dilis, V. (2006), 'Traditional foods: Why and how to sustain them', *Trends in Food Science & Technology*, vol.17, pp. 498–504.

Van den Steen, J. (2006), Geuze en kriek: de champagne onder de bieren, Leuven, Davidsfonds.

Vaz, T. N. and Nicolas, F. (2000), 'Innovation in small firms and dynamics of local development', Workshop ISEG, 24–25 January, Lisbon, p. 11.

Websites

http://ec.europa.eu/agriculture/foodqual

Chapter 15

Protected Designation of Origin, Sustainable Development and International Policies: A Survey of DOC Wines from Emilia-Romagna

Silvia Gatti

15.1 Sustainable development

Today, the concept of sustainability significantly affects our life and economic activities. In 1987, the Bruntland Commission stated: 'Humanity has the possibility of making development sustainable, i.e. ensuring that the needs of the present are met without compromising the ability of future generations to meet their own needs' (World Commission for Environment and Development, 1987).

By means of the 6th Framework Programme for Environment, environmental sustainability was adopted as a mark for all the policies the European Union. Today, the concept of environmental sustainability is applied to agriculture to all intents and purposes through the cross-compliance of the Medium Term CAP Reform and the Good Farming Practice (GFP).

Furthermore, Kyoto Protocol detects two gases responsible for the greenhouse effect which mainly originate from agricultural activities: CH_4 Methane gas (the emissions of this gas originate from agriculture, animal faeces, and waste dumps), and N_2O Nitrogen protoxide (agriculture, energy sector and transport are also responsible for this gas).

15.2 The international safeguard of the Protected Designations of Origin

This study originated from the perception that, in such a crucial stage of both international negotiations and EU politics, the safeguarding of one of the most important assets of regional agriculture must be reinforced, by connecting the links of the designation of origin to the characteristics of the ecosystem and the economic and social fabric.

In his book describing the path of implementation of the Italian Act No. 164 on designations of origin (Fregoni, 1994), Professor Mario Fregoni states that 'in order to understand the essence and definition of a designation of origin it is necessary to recall that wine's organoleptic characteristics are correlated to the *vine's ecosystem:* this is made of vines and primer pouches, climate, soil and human techniques, biological- ecological- and anthropoganic- supports of the designations of origin' (Fregoni, 1994, p. 100). In this perspective, the relationship between wine making and vine-growing with Protected Designations of Origin (Denominazioni d'Origine Controllata, DOC) for Italy, in Emilia-Romagna, and the actions aiming at Sustainable Development were investigated.

The initially defined investigation areas were those of Colli Piacentini DOC wines, Colli Bolognesi DOC wines, as well as Sangiovese di Romagna DOC wine and Albana di Romagna (DOCG – Denominazione di Origine Controllata e Garantita) wine. The remarkable presence of production characterized by DOC in some flatland areas, and the connection between the policy of sustainability and the biggest cooperative enterprises induced us to extend our study also to the areas of Modena and Reggio Emilia as well.

This work studied the actions that have been carried out in the wine sector of Emilia-Romagna to achieve environmental sustainability. For this purpose the study used the available data on voluntary certification, the variety and extent of autochthonous vines, as well as the data concerning organic farms and enterprises.

15.3 Environmental damage caused by the vine-growing and wine-producing sector in Emilia-Romagna

Paolo Neri (2005), from ENEA, on the determination of the environmental damage attributable to the production of a 0.75-litre bottle of wine made of conventional grapes showed that, according to the Ecoindicator 99 E and EPS 2000 evaluation methods, the most damaging stages are the implementation of the vineyard, pressing and fermentation, stabilization, cultivation and transport. In particular, the EPS 2000 evaluation method shows damage equal to 2.57 ELU (Pt) of which 36.5 per cent is due to cultivation stages, 30.6 per cent to pressing and fermentation, and 26.1 per cent to stabilization. The damage affects: Human health (7.3 per cent), Ecosystem production capacity (43.6 per cent), Abiotic stock resources (49 per cent), and Biodiversity (0.14 per cent).

In the light of these results, the environmental damage caused by wine making seems to be rather equally divided between the agricultural stages and the industrial stages in the process. According to the Ecoindicator 99 method, the loss of biodiversity due to the occupation of the land by the vineyards is especially relevant.

A comparison of the analysis carried out at the same time for Prosciutto di Parma [Parma Ham] (damage equal to 11.5 Pt) and Parmigiano Reggiano (damage equal to 16.7 Pt) shows that winemaking (in this case Sangiovese) causes less environmental damage if compared with the life cycle of these other relevant quality agro-food products. Even more striking is the difference with industrial production such as sugar beet (damage equal to 26.4 Pt).

15.4 Participation in ISO 14001 and EMAS certification systems for the vine-growing and wine-producing sector in Emilia-Romagna

In order to assess the approach to the management of environmental problems by the wine sector in Emilia-Romagna, the number of ISO 14001 and EMAS certifications was determined. UNI EN ISO 14000 regulations date back to September 1996; they define the requirements of an Environmental Management System (EMS SGA), while taking into account both legal requirements and information concerning significant impacts. This is a voluntary system.

The EMAS (Environmental Management and Audit Scheme) regulations, originally issued in 1993, solely addressed the industrial sector. Only at a second stage, in 2001, was its application also extended to the service sector, and, in general, to every kind of organization. Unlike UNI EN ISO 14001, EMAS is recognized at the European level: it seeks a voluntary engagement with environmental safeguards, by reorganizing and rationalizing the environmental management of the member organization, which will be based not only on respect for the legal limitations but also on a new relationship involving the organization itself, institutions, the industrial world, and the general public.

As of November 2006, Emilia-Romagna was the 4th region in terms of number of UNI EN ISO 14001 certifications (890), after Lombardy (1332), Piedmont (1017) and Campania (891). As for EMAS, our Region is first in terms of certification numbers, exceeding by far those registered in Lombardy, Veneto and Piedmont.

At the national level for the 'food- drink- and tobacco- industries' sector, 667 ISO 14001 certified sites are registered (125 certifications), about 83 of which are in the winemaking sector. In contrast, for the 'agriculture and fishing' sector, 67 certifications can be found at the domestic level, only six of which are in the vine growing sector (data collected in November 2006). Emilia-Romagna has 14 out of the 45 ISO 14001 certified sites in Italy for the winemaking sector.

The 14 ISO 14001 certifications include enterprises and most of the cooperatives. ISO 14001 certifications (combined with EMAS certifications) resulted from the professional standards achieved by the first large cooperatives. The professionals who certified Parmigiano Reggiano dairies are presently working for the certification of wine growers' associations outside this region.

There are 4 EMAS certified enterprises in the wine sector in Emilia-Romagna. In Italy as a whole there are only 6 EMAS certified winemaking enterprises. Overall, the EMAS certified agro-food industries were 89 as of 2006.

15.4.1 Certification costs

The expenditures for the implementation and maintenance of an Environmental Management System can be classified as: a) expenditures for the introduction of the system; b) expenditures for the certification of the system; and c) expenditures for maintaining the system.

According to a study by CESQA of Padua University carried out in cooperation with SINCERT on organizations active in Italy and ISO 14001 certified in 2001, for SMEs these costs are slightly lower than € 25,000 whereas for large enterprises they slightly exceed € 45,000 (plus the expenditures for external advisors) (Provincia di Asti, 2004).

It must however be underlined that some of the largest enterprises which disregarded the strategy of environmental certification did in fact invest in traceability systems.

The procedures for the management of environmental problems by winemaking enterprises which have not adopted environmental certification systems are governed by the relevant regulations in force

15.4.2 Competition in certification

Environmental certification allows the enterprises to be in the market with their papers in order from the environmental point of view and, thus, to increase the operators' confidence. The survey on ISO 14001 or EMAS certified enterprises underlined the strong connection between the choice of environmental certification and the need to meet the standards set out for access to public funds.

The same enterprises stressed that they had decided to comply with other standards required by European large volume distribution certified at their own cost. In particular, the Austrian, Swiss, French and German chains require the IFS (International Food Standard), while British chains require the BRC[1] standard. Furthermore, presently, consumer cooperatives in Italy require the SA 8000 ethical standard concerning the respect for human rights, the respect for workers' rights, the prevention of the exploitation of non-specialized workers, and the safeguard of occupational health and safety.

15.4.3 Ecolabelling

For agro-food products the possibility of ecolabelling has not been set out. However, there is the QC (*Qualità controllata*) mark linked to integrated pest management production organized by Emilia-Romagna. It represents a corporate policy, for instance, for the entire system of associated wine cooperative and vine producers (i.e. 90 per cent of the associated producers in wine cooperatives have joined in the project of integrated pest management production).

---

1 Developed by distribution chains, they have the purpose of enhancing the effective selection of own brand food suppliers for large volume distribution on the ground of their capability to provide safe products which meet both contract specifications and legal requirements. This standard indicates the specific factors of a management system focused on quality and hygiene and health and safety of the processes; this system uses the HACCP method as a reference for planning and implementation.

15.4.4 Environmental certification and quality production

The results of the survey underline the close link between large enterprises in the wine sector and environmental certification. Another important factor is the mainly cooperative nature of the enterprises which follow this policy. In the environmental certification of one of the most important wine making cooperatives in Emilia-Romagna, it is evident that the choice to adopt a policy of control over the environmental impact also originates from both the cooperative nature of the company and its strong link with its territory and population (Cantine Cooperative Riunite, 2005).

Nevertheless, a strong link between the policy of DOC quality and environmental certification does not seem to exist. The *Consorzi di Tutela* (Protection Bodies) do not seem to provide incentives for specific behaviour in this domain. The top DOC wine producers in Emilia-Romagna are not among the certified companies.

15.5 Preservation of biodiversity

In the past 30 years the populations of land species have decreased on average by 15 per cent, marine species by 35 percent and fresh water species by 54 per cent (WWF, 2004).

The work by Paolo Neri (2005) shows that the occupation of the soil for vine cultivation represents a remarkable loss of biodiversity. The uncultivated soil would have the possibility of hosting many more animal and vegetable species. However, vine growing in Emilia-Romagna also has the role of safeguarding and stressing the value of biodiversity through the recovery of autochthonous vine varieties and of long-standing cultivation. Here these varieties are used as a marker of sensitivity for the more complex problem of biodiversity, as well as a factor of growth for the policy of designations of origin.

15.5.1 Autochthonous vine varieties

International vine varieties have an important role in Italian DOC production and in that from Emilia-Romagna in particular, and the safeguarding and valorization of autochthonous vine varieties are not explicitly mentioned in Act No. 164/92 on wines' designations neither of origin nor in the regulations for DOC wines in Emilia-Romagna. But the Rural Development Plan of the Emilia-Romagna Region does include the development of autochthonous varieties of pear and apple trees and vines amongst its provisions.

The agreement reached on 10 February 2005 amongst MIPAF (Ministry of Agriculture and Forests), autonomous Regions and Provinces, concerning the 'Safeguard and valorization of productions obtained by either autochthonous vine varieties or vine varieties of old standing cultivation', sets out that the use of the designation of the vine variety and its synonyms on the label may be subject to 'limitations' in consideration of the origin and the strong link of the vine variety itself with the territory, the typicalness of the obtained production and historical-cultural aspects, obviously bearing in mind the effective spreading of the vine variety.

In 2000, the 5th General Census on agriculture showed that 50.7 per cent of the overall area of the main vines in the Emilia-Romagna Region was occupied by autochthonous vine varieties, while the incidence of autochthonous vine varieties for the areas devoted to DOC and DOCG wine production was slightly greater, reaching 55.3 per cent.

For international vine varieties, the presence of Chardonnay, Sauvignon and Cabernet is mainly concentrated inside the DOC and DOCG areas and reaches 4.9 per cent, while the Merlot vine (4.0 per cent of overall area) is largely present in the areas devoted to table and IGT (*Indicazione Geografica Tipica – Typical Geographic Indication)* wines.

Substantially, all the regulations of hilly region DOC include one or more wines with an international vine variety name. Lambruschi DOC and Reno DOC areas are excluded. In addition, the latest regulation of 'Colli di Romagna Centrale' of 2001 sets out wines with international vine variety names.

15.6 Organic cultivation in the vine-growing and winemaking sector in the Emilia-Romagna region

15.6.1 Statistical basis

Out of the overall 4084 farming enterprises and 681 organic transformation enterprises in Emilia-Romagna, 723 include the production of grapes in their production range as of 31 December 2003, according to the Official List of Organic Operators of Emilia-Romagna (it must be underlined, however, that some of these enterprises did not declare their production range). 48 organic enterprises were certified for grape transformation (it must be emphasized again that only the enterprises which stated their production activities and, specifically, grape transformation were taken into account).

In order to assess the weight of the organic producers in the domain of quality vine-growing wine-production, it was investigated how many of them had also joined in *Consorzi di Tutela*. Then, the list of organic operators was compared with the lists of members of the *Consorzi di Tutela* of the 3 DOC and DOCG wines examined. This was a difficult task because, very often, the same enterprise appeared with different names in the two lists. Therefore, our results must be considered to be more indicative than absolute. Keeping this in mind, we can state that the organic enterprises which joined in were from:

- the 'DOC Colli Piacentini' Consortium: 7 out of 80 members, all of which are vine and wine producers;

- the 'DOC Colli Bolognesi' Consortium: 13 out of 170 members, of which 8 are vine and wine producers and bottlers out of 62, and 5 are grape producers out of 85;
- the 'Ente Tutela Vini Romagnoli': 9 out of 83 members.

15.6.2 Regulations

Wine is explicitly excluded by Reg. No. 2092/91. Therefore one cannot refer to 'organic wine' or 'wine obtained with organic procedure' in the framework of the regime defined by the European regulations which do, however, permit the statement 'wine obtained from organic grapes'. This is because the 'farming' stage is governed by the above- mentioned regulations, whereas the transformation stage is not. Why this stage is not governed by European regulations is not clear. Cookies, juices, cheese, and other products are.

A consequence of this uncertainty is linked to the (voluntary) use of the European brand. In Germany, England and Austria 'organic wine' or 'Ökologish wein' labels are used, while in the extra-EU countries that are importing considerable quantities of wine – such as the U.S. and Japan – an 'organic' statement on the label is an essential prerequisite.

For some time, the organic producers' associations in different European countries have self-imposed their own regulations for winemaking: regulations which are voluntary and not binding for everyone (Micheloni, 2002).

15.6.3 Disciplinary measures and sulphur dioxide

A criticism of the production of organic wine concerns the problem of the management of wine stabilization and, thus, the use of sulphur dioxide. The CCPB disciplinary measures (2003) show the possible use of a small quantity of acidic oxide (see Table 15.1).

Table 15.1 CCPB acceptable acidic oxide levels in organic wine

Type of wine	Recommended SO_2 values (mg/l)	Admitted SO_2 values (mg/l)
Red wines	< 20	80
White and rosé wines	< 20	90
Sparkling wines	< 20	60
Sweet and sparkling wines	< 20	100

Source: CCPB Disciplinary Measures (2003).

The AIAB disciplinary measures (Micheloni, 2002) set out total recommended SO_2 values as shown in Table 15.2.

Table 15.2 AIAB acceptable SO_2 levels in organic wine

Type of wine	Recommended SO_2 values (mg/l)	Admitted SO_2 values (mg/l)
Red wines	< 20	60
White and rosé wines	< 20	80
Sparkling wines	< 20	60
Sweet wines	< 20	120

Source: AIAB Disciplinary Measures (Micheloni, 2002).

15.6.4 The sustainability of organic production

According to studies carried out by Della Giovampaola and Neri (2004) and Neri (2005), the comparison between conventional and organic wine shows that organic production has an environmental impact 7.8 per cent lower if compared with conventional production (according to the Ecoindicator 99 method). In particular, it appears that when passing from conventional to organic production, the harm to Human Health is reduced by 9 per cent as a result of two opposing phenomena: 1) reduced transportation from production to consumers which, in turn, induces a lower production of SO_2 and CO_2; 2) the use of the facilities for a smaller quantity of wine, thus inducing a greater production of NO_X and N_i. As for Ecosystem Quality, the negative impact increases by 3 per cent as a consequence of the land use. This occurs because, even though there is less intensive exploitation of the soil, in organic cultivation the damage registered is greater on account of the smaller production of grapes on the same area of land. As far as Resources are concerned, the damage is decreased by 24.4 per cent mainly thanks to: 1) transport from producer to consumer only in Italian territory, for organic products; 2) 20.4 per cent reduction in energy used, especially thanks to decreased transportation; 3) 19.3 per cent decrease in expenditures. The assessment according to the EPS 2000 procedure shows that organic production causes 5.4 per cent less damage than conventional production.

A problem in organic vine and wine production concerns the use of copper for treatments. This causes very strong pollution of the soil thus requiring limitations in its use.

15.6.5 Does an Organic-DOC production make any sense?

One reason for the scepticism about the idea of Organic-DOC products – that is the creation of a niche of organic DOC products to be recognized and organized by the *Consorzi di Tutela* of DOC products in cooperation with organic consortia – is that organic production is much more exacting than DOC production. According to a rather widespread opinion, in organic production the certification concerns both the plants – with the need, at present, for an independent transformation line

– and production. Therefore, the participation in the management of such a project by the *Consorzi di tutela* would not be easy to put into practice.

The *Consorzi di tutela* bring together very different types of enterprise and, therefore, do not deem it advisable to pursue this aspect. Thus, organic DOC producers are organized in two consortia, the activities of which are not specifically coordinated. The difficult integration of their policies has sometimes caused the enterprises which decided to go organic to leave DOC and join in IGT.

Big enterprises such as cooperatives have Organic-DOC production. For the time being, the organic Lambrusco Grasparossa DOC wine is produced in small quantities (20,000 bottles) for the German market. Its bottle has the same price as the top lines in large scale retail distribution. The organic project of cooperatives in the Lambrusco area faces difficulties concerning the characteristics of organic enterprises: 30 farms produce organic grapes, but only 5–6 of those are in the most interesting area for cooperatives: the hilly region.

The need to export all over the world requires, in addition to the 'European' certification, also the US NOP (Natural Organic Programme). In Emilia-Romagna, organic winemaking seems to have the possibility to grow if: a) organic wine can be sold directly to the consumer; and b) organic production can contribute to the positive image of the entire farm project, and especially of the agro-tourism activity.

15.7 Sustainability of DOC wine making

Act No.164 of 1992 states in Section 1 that: 'The designation of origin of a wine indicates the geographical name of an area especially devoted to vines; it is used to indicate a well-known quality product, the characteristics of which are connected to both the natural environment and human factors'.

Section 10 defines the contents of the disciplinary measures for production. Item f) states that: 'The conditions of production and, especially, the natural characteristics of the environment, such as: climate, soil, location, altitude, exposure, ampelographic composition of the vineyards devoted to the production of grapes belonging to recommended and authorized vines varieties, plantation density, breeding forms, pruning systems, and forbidden forcing practices' shall be defined.

In the revision of the policy concerning the DOC wines, under this item f), some prescriptions for sustainability could be included. Thus an environmentally 'virtuous' behaviour of DOC production could be defined.

Professor Fregoni's concept of the vine ecosystem, on which the idea of wines' designation of origin rests, can be met and brought up-to-date by paying greater attention to the product quality, its bio-diversity characteristics and their safeguarding, as well as to the agronomical and transformation practices. This means that the carrying capacity of this ecosystem is not exceeded, for instance, by controlling the contents of plant protection products used at different production stages.

In Emilia-Romagna, the relationship between DOC and sustainability can be found in the disciplinary measures for integrated pest management production. In these, as well as in the Rural Development Plan, attention is paid to DOC production.

15.8 Conclusion

The starting point for this study was to understand the relationships between the new environmental sustainability sensitivity, international policies, and designations of origin.

The work, carried out in Italy on the most important vine-growing and winemaking areas of Emilia-Romagna region, required, first of all, the determination of some indicators of this sustainability. Therefore, the studies on the analysis of the life cycle of wine were considered, comparing them with the region's other important quality agro-food production and with industrial production. The ISO 14001 and EMAS environmental certifications were considered for vine-growing and winemaking, and also the role of the QC (Qualità Controllata) mark was taken into account for the integrated pest management production in Emilia-Romagna. Lastly, the organic production activity for vine-growing and winemaking was analysed.

The analysis of these indicators and the interviews conducted with privileged contacts showed how, beyond the definition of 'designation of origin' itself, which centres on the vine-growing ecosystem, the sensitivity to environmental problems is managed on two levels: at the vine-growing enterprise level, and at the level of *Consorzi di tutela* and the local authorities.

The vine-growing and winemaking enterprises of Emilia-Romagna for the production of denomination of origin wines covered in the study have different behaviours depending on size. Organic practices are found amongst the small quality enterprises. Amongst the large enterprises, organic production lines can be found, but mainly this is where environmental certifications are concentrated. For their part, the *Consorzi di tutela* and regional research centres have focused particularly on the development of autochthonous vine varieties, although the presence of international vine varieties is still very strong in the region's regulatory measures.

It may thus be concluded that the development of an environmental sustainability policy for designations of origin, in keeping with the provisions of international policies, is growing partly from the bottom-up in Emilia-Romagna, through the initiative of the single sector operators, and partly through the *Consorzi,* and especially, the local authority, focusing on several specific actions. At the present time, the only point where the product and territory policies of the designations of origin cross paths organically with environmental policies is in the Region's Rural Development Plan.

It is the writer's opinion that a policy of environmental sustainability of designations of origin would be a strong point for their safeguarding in the international bargaining on free trade, by bringing the product-territory-environmental protection link to a privileged level over that of simple brand name logic.

References

Cantine Cooperative Riunite (2005), *Dichiarazione ambientale Cantine Riunite,* at http://www.riunite.it/ita-home.htm

CCPB (2003), *Controllo biologico, Norme di produzione biologica*, Edizione 02, gennaio.

Della Giovampaola, M. and Neri, P. (2004), 'Sangiovese "Bio", anche l'ambiente ci guadagna', *Agricoltura*, settembre.

Fregoni, M. (1994), *La piramide DOC. Storia, contenuti, interpretazioni della 164/92*, Bologna, Edagricole.

Micheloni, C. (2002), *La viticoltura biologica italiana*, at http://www.aiab.it

Neri, P. (2005), *Analisi del ciclo di vita del vino Sangiovese, Confronto tra produzione convenzionale e biologica,* L'analisi del ciclo di vita (LCA) dei prodotti agro-alimentari, Bologna, 21 febbraio.

Provincia di Asti (2004), Introduzione ai sistemi di gestione ambientale nelle aziende viticole, Asti.

World Commission for Environment and Development (1987), *Our Common Future,* UNEP.

WWF (2004), Living Planet Report, Switzerland, Gland.

Chapter 16
Large-Scale Production and Market Segmentation: An Uneasy Relationship

Teresa de Noronha Vaz and Peter Nijkamp

16.1 The context

A glance at the current state of the food processing industry reveals a perspective of restless and energetic activities adjusting to shifting consumer desires and global markets. Modern society has managed to adequately shape the supply of and demand for industrial food products in a way that encourages us to assert that the produced amounts of goods are sufficient. However, the huge marketing and logistic efforts put into these activities are not yet able to provide processed food in an equitable and balanced way for the entire world's population. The reasons for this situation concern both marketing channels and income distribution.

In fact, international food provision, as we may call it, has been determined by approaches which have stressed, over the years, globalizing tendencies. The established political economy of the agro-food systems favours multinational food processing and retail capital which encourage free trade and justify a global view. But there is a danger that small firms may be excluded from the agro industrial system.

Connor and Schiek (1997) made a very accurate study of the evolution of the food processing industry and distribution for the United States, pointing out not only the determinants of food expenditures but the steady evolution of the industrial structure and business organization in the food sector, up to about 1995. For the USA, the country that is the leading example of food consumption in the developed world, the authors detected what they called a 'constancy of change'. Characterized by very slight alterations in consumption patterns and by strong concentration strategies of productivity gains, such an apparent slow motion progression would, however, generate conditions for significant change in the long run. Connor and Schiek saw this as a 'powerhouse under reconstruction'.

In spite of the concentrating tendencies and the increasing globalization or, probably, as its consequence, the 1990s brought to the socio-economic scene new developments for the industrialization process of food processing. Derived directly from farmers and consumers, attempts have been made in order to challenge the global agro-food complex and propose other, alternative, solutions which introduce new forms for the existing food systems. There are different types of consumer concerns to justify such attitudes: 1) food safety and human health; 2) the long-term environmental impacts of industrialized agriculture; 3) agricultural welfare; 4) animal use; and, finally, 5) issues associated with fair trade at a world scale.

Notwithstanding this empirical approach, the scope of this discussion goes far beyond the context of food production alone. Indeed, during the last two decades, several authors[1] have put forward serious scientific arguments that there is a crisis of global capitalism, which Soros (1998) saw as a danger resulting from market fundamentalism and the resulting attempt to replace collective decision making with market values overriding all social and political considerations.

Although we do not pursue the discussion about the ethical constraints of market relations, it is still hard to ignore Soros's arguments when dealing with an industrial activity for which unstoppable expansion is mainly determined by: multinational industrial companies linked to research laboratories; marketing channels basically controlled by powerful international wholesalers; and vertically integrated, or coordinated, influential financial groups.

Along with this view, a strong reaction to the state of affairs in the modern food systems can be observed in Bodin-Roudier's (2000), book entitled 'La guerre alimentaire a commencé' in which the author dedicates the final pages to an open letter addressed to the world's biotechnology firms. It is an emotional text, probably too *engagé* and in some parts exaggerated, calling attention to the need to reflect carefully about what is taking place within the present structure of the food industry and how little from our past habits and food tastes may be preserved. What is under discussion in the author's arguments and what we would like to stress here is whether a past cultural food heritage related to many social and economic diversified structures, most of them with local identities, should be given up for the sake of a global industrial model for which long term advantages may not be safe or fair.

Much has been done since the 1980s, when the heavy Common Agricultural Policy (CAP) budgets were worrying European policy makers and international market leaders menaced the success of Common Market Organizations (CMOs). But there was no other choice for a sector which has changed the post-war social and economic aspects of Europe: agriculture and the agro-food industry, in general, created the basis for growth and development in most European countries; they guaranteed much income in rural areas, sustaining farmers and small towns. Nevertheless, today, the data confirm the existence of 19 million jobs in the food-sector representing 9% of total direct employment. This does not include the indirect push that the food industry has given to technology, supporting it with research funding, new ideas, and demands for new products. Much of the product innovation in the European context is of an incremental nature, and its provenance is from the agricultural and food sectors.

These arguments are of crucial value when considering the issue raised in this chapter from the perspective of the social nature of rural development. It is not possible to discuss the costs and advantages of both large-scale production and

1 For a more complete bibliographic support on this issue, see Patnaik, 2003.

market-segmentation productive models without considering the respective long-term and cross-sector effects. This is probably the reason why, for example, the programme LEADER – a very successful instrument of local development in rural areas – is now considered an integral element of the CAP, which is budgeted for € 88 billion for the period 2007 to 2013, which represents 21 percent of the total spending under the CAP. Independently managed, this fund opens new prospects for local environments as it may improve services and quality of life, in general, and support a diversity of rural economies. Past experiences show that this programme facilitated the creation of new networking systems and partnerships which, in general, represent indirect supports to market segmentation productive models.

16.2 The evidence

16.2.1 Theoretical

Market power There is no doubt that the economic performance of large firms in the food manufacturing and retail distribution industries is critical to the overall performance of the food sector. During the 1980s, the concentration of companies led to such profitability gains that food firms' stock prices increased by about 900 percent (Cotterill and Iton, 1993). Those companies became interesting research objects and the respective case studies fascinated a multitude of scientists and strategic analysts. They wanted to find the reasons for these exploding values in firms' concentration forms, market share distributions, cost determinants, size effects, companies mobility choices, price strategies, or public policies.

A major set of studies to theoretically explain this sequence of profits for firms in the food sector resulted from Weiss's (1974) insights on market power and its consequences for industrial organization. In addition, some scientific material related to the Profit Impact of Market Strategy (PIMS) confirmed that '*the positive profit-market share relationships observed in line of business studies represented a still- unknown mixture of temporary efficiency differences and more a less durable monopoly power*' (Scherer and Ross, 1990)

Amongst others (starting with the pioneering paper of Schrader and Collins, 1960), the study carried out by Cotterill (1993) developed an analysis of the structure-performance relationships within the food systems. The results support the following conclusions: 1) that concentration is a major source of profitability at the business unit level, and this not because of aggregation bias; 2) that share or relative share are positively related to profitability, and this not caused by lower costs; and, thereby, 3) that market share and concentration persist as indicators of market power in the food manufacturing industries.

In spite of the relevance of concentration and integration for the development of the food industry and distribution, nevertheless, other 'residual' industrial formats have survived the dominant oligopolistic structures. Meanwhile, constraints to

Taylorism justified the emergence of new productive models considered to be more suitable for growth in Europe and other parts of the world, particularly when facing the requirements imposed by the indispensable flexibilities of firms. Boyer and Durand (1998) clarified how the industrial mass production model, dominant for more then 50 years, proved to be inadequate, for reasons of conditions of permanent technical change, growing consumer demands, and tight labour regulations. Hence that model started to give way to other, more efficient, models supported by segmented productive processes.

Though many theoretical essays have been written in this regard and empirical observation corroborates the importance of a new phenomenon, one may suspect that a certain level of anxiety has puffed up the speculative arguments for the end of the Fordist/Taylorist era. Nonetheless, flexible production has indeed spread as a new productive form, imposing recent changes in the national regulations related to labour and transaction costs, as clearly discussed in Boyer and Saillard (1995). Vaz (2007) indicates a possible intermediate position, suggesting that a permanent but non-synchronized movement towards segmentation and flexible specialization is taking place within the agro-food sector along with an unchangeable structure led by large corporations.

Also, as a part of this background, Coriat and Dosi (1995) provided theoretical tools to analyse the impacts of economic change due to technological innovation, in order to reconcile the evolutionary approach with regulation. In the authors' opinion, the search for the roots of economic dynamics lies in the imperfect rationality of all the agents supported by routines and learning – which, in our opinion, are forms that both encourage diversity and maintain the track traced by path dependencies.

Providing evidence for the flourishing of combined forms of labour specialization in a context of local identity, the set of arguments supplied so far has opened a theoretical framework for alternative industrial forms (if not yet models) based upon segmentation and networking, thereby constructing the most robust argument to sustain the value of traditional production in the food system. Labour specialization, combined with long-term routinized productive forms, integrates learning and permits innovation at a time when the easy incorporation of labour and social path dependencies may become crucial in the sustainability process of rural areas. The impact of such change in a new geography of food, as stated by Parrot et al. (2002) is evident.

The particular case of traditional food production serves as a unique example of structural change, because it surpasses the simple productive model to combine social goals of fair distribution and trade. Areas allocated to traditional food production, all over the world, need to be kept competitive – a market imposition which goes along with the acceptance of new learning processes. Only integrating technological and organizational innovation can bring to the traditional food industries the required firm efficiency (Vindigni et al., 2006; Copus and Skuras, 2006) and new opportunities to promote rural and local growth. However, because of the rigidity of many micro and small firms in charge of traditional food production, the updating of productive techniques and the necessary modernization of organizational forms represent a hard way to go (Vaz et al., 2006). This is so, independently of the major role these firms may represent in the socio-economic transformation of the areas in which they are rooted. Studies on firm embeddedness in the food sector (Avermaete et al., 2004) show how valuable the good performance of small firms in the rural areas could be for local development.

From vertical coordination to corporate social responsibility The tendency for desegregation of the productive processes imposes practices of shared information, in order to reduce uncertainty and risk for consumers and producers, but also entangles the structures of food manufacturing. In order to assure consumers, product traceability (back from the farm upstream to the distribution agents) is emerging as a precious instrument to guarantee food safety and quality, even if additional costs are foreseen and permanent communication within the food chain has to take place. Vertical coordination provides the context for many different actions (Galizzi and Venturini, 1999). Examples are the public-private partnerships to discuss and regulate food safety, the importance of vertical coordination to calculate risk assessment, or studies on food purchasing behaviour.

The Supply Chain Management (SCM) concept was developed in order to find the optimal co-ordination forms able to reduce costs, in general, but also to increase effectiveness and add while reducing uncertainty and risk. But the food industry was slow to follow the principles of SCM, although there has been considerable scientific research dealing with the different forms to progress in this area.[2] As competition for market share strengthens in the food retailing industry, concentration and brand loyalty becomes increasingly important for food manufacturers and retailers. Specific methods of market segmentation can be found in order to develop brand loyalty amongst distinct consumer groups.

Given the concern about the composition, origin and environmental consequences associated the food acquisition, it is not surprising that a natural reaction for consumers might be their readiness to alter their consumption habits (Guptill and Wilkins, 2002) – attempts have involved educational campaigns to raise awareness of dietary priorities and encourage the consumption of healthier foods. In the process, consumers are becoming conscious of the extended impact of their behaviour upon how food production, distribution, manufacturing and retailing take place and the discussion of strategic choice in the classical Taylorist/Fordist model has acquired a new ethical perspective.[3] On the other hand, in exploring the best practices regarding, for example, waste management, sustainable sourcing or supply chain relationships, large corporations have much to offer. In brief, more

---

2 Efficient Consumer Response (ECR) and Collaborative Planning Forecasting and Replenishment (CPFR).

3 For example, criticisms of life science companies are intensifying as they react to European concerns about genetically modified crops moving to less developed regions of the world.

research to help in the calculation of the positive and the negative effects of both models upon the global food chain is of the utmost urgency.

16.2.2 Empirical

Market segmentation and labelling The latest trends of American corn production provide an illustrative case for market segmentation across the world. This crop, being the most commonly produced feed grain in the United States, represents more than 90 per cent of the total value and production of feed grains and occupies roughly 80 million acres of land, mainly located in the Heartland region. The product is mostly used for livestock feed or is processed into different food ingredients (starch, sweeteners, corn oil, beverage and industrial alcohol) or fuel ethanol. In the international scene, the USA represents the most important actor in the world corn trade market, exporting about 20 per cent of its corn production. There is no better commodity to exemplify the advantages of large-scale production. However, even in this case, markets are demanding different specific product identifications, which requires adjustments along the value chain.

Given the importance of corn as an agricultural export commodity and also a raw material for American industry, in 2003, the Economic Research Service (ERS) of the US Department of Agriculture,[4] together with the US Farm Foundation organized a conference on Product Differentiation and Market Segmentation in Grains and Oilseeds. This meeting explored the determinants of market segmentation and the implications for growers, handlers, processors, and consumers. The variety of the presented by-products and the arguments in favour of incorporating specificities and local identities as added values in the supply chain were generalized contributions and should serve to draw our attention to a shift in the large-scale productive model and the corporate attitude towards mass production.

Furthermore, the clearest manifestation of the explosion of diversity within food production is reflected in the implementation of new private labels, aiming to assure consumers of higher quality standards and more food safety. Labelling has a sequence effect upon the whole supply chain, which is reflected in the increasing specifications in the long-term contracts with upstream producers. For example, Bazoche et al. (2005) have analysed, for the case of France, the interest of producers to commit to such private labels, their effects on spot market prices, and the resulting market segmentation between the spot market and supply contracts for meat and fresh vegetables.

Alternative agri-food initiatives In a much less organized and disperse form but frontally challenging the premises of concentration and mass production, another significant number of diverse empirical examples can be supplied by the literature

4 ERS is in charge of developing permanent research programmes to help public and private decision making on food, farming, natural resources, and rural development.

review which has identified them as 'alternative agri-food initiatives' (AFIs).[5] All this evidence provides examples of emerging structures and includes community-supported agriculture, farmers' markets, urban agriculture, and regional food labels.

In practise, these efforts stand for a new kind of activism that links consumers to the traditional focus on farmers and production. In this new politics of food, activism is embodied in quotidian necessities and experiences, representing commitments able to join and motivate different interests and people throughout the whole food chain. '*Accordingly, AFIs are increasingly celebrated in both popular culture and academic venues as agents of progressive change in the agro-food system. As compared with the conventional food system, they are considered to be more equitable and environmentally sound, as well as providing healthier food*' (Allen and Goodman, 2001).

All over Europe and the USA there are many examples of AFIs to be reported. We shall present just two of such cases in order to bring awareness of an emerging phenomenon. Our first selected case reports a prosperous range of alternative agro-food initiatives in South West Ireland. Specialized in producing high value 'gastronomic' foods, lifestyle or 'alternative value' products, the initiatives integrate their own distribution channels. Chain coordination is given by a multiplicity of overlapping networks, whose synergies bring together different specialists occupied in the manufacture and supply of high quality products for consumers of different profiles (traditional food production included). In this case, there is an organizational effort which is reflected in the product brands rather than in producer identities – they represent an ecological and spatial origin which stands for product quality. *Producers, particularly the leading ones, may be endowed by consumers with a measure of trust, or moral authority, which allows them to speak out, for example, in defence of traditional production methods, in opposition to state-imposed hygiene regulations*[6] (Murdoch et al., 2000).

In the same region a different network, but with similar priorities, can be identified. It connects farmers, growers and other food producers wishing to sell directly to consumers. This case of embeddedness is mostly caused by the proximity of production and consumption but represents a reshaping of the long-term established commercial channels for foods. The new forms of exchange are promoting complex social relations, which although hybrid could be able to sustain many alternative agro-food networks (Sage, 2001).

As a matter of fact, worldwide web initiatives are easing the diffusion of this kind of AFIs, by promoting direct relationships between food producers and consumers. Many informal marketing structures like retail farmers' markets or community-supported agriculture ventures are making good use of Information and Communication Technologies (ICTs) to publicize alternative marketing channels and

---

5 We could include traditional food production in such a classification.

6 Which particularly have an impact on the small-scale food sector or on consumers' interests (e.g. GM food).

change the existing long-term contracts which for decades have shrunk the market power of producers, as proximity of production gave way to global marketing.

Our next identified case is located in Iowa, USA, a state dominated by conventional commodity agriculture for corn, soybeans and hogs and mainly dependent on food imports. According to Hinrichs (2001), Iowa provides a useful example of politics connected to 'local' foods. It seems that a growing 'local' foods movement may be observed but is associated with conflicting attitudes. Although promoting significant producer and consumer opposition to conventional commodity agriculture and industrial foods, this movement tends to '*assume outcomes of local social and environmental relations that may unwittingly feed nativist sentiments and highlight selected environmental achievements, while ignoring more subtle failures*'.

16.3 Concluding remarks

In exploring the potential importance of the renewed role of segmented markets, we identified the exemplar case of traditional food production, but we neglected a major function of the small-scale firms of this sector. As they are employers of a great number of people across the world, we can not escape their links to social responsibility evolving the previous considerations much further than a profit-performance argument. What we are dealing with comprises a multiple task of technological change, social efficiency, and environmental responsibility that needs to be achieved by any future industrial model for the agricultural systems.

By raising awareness of the capacity of such low technological regimes, it might be possible for them to benefit from technological change, sometimes even introducing radical innovations and, consequently, also modifying the respective socio-environmental contexts. There are social implications that can be studied, particularly those that reorient social needs towards consumption patterns centred in quality standards. In addition, the context for innovation in small, fragile firms is a very interesting area of research: the development of firm-strategic alternatives for niches as a tool for sustainable development (Smith, 2006). In a typical environment of learning regions and of segmentation of productive activities in the food systems, we should recognise that peripheral Europe has the prerequisites for technological apprenticeship in its historic-cultural past. Furthermore, within a territorial space, collective actions and organized sectorial interdependencies perform functions such as research, selection, codification, transformation, control and other procedures that, in the end, represent constructive knowledge flows able to transmit modern know-how to low, as well as to high, technological sectors.

In the case, for example, of the segment related to traditional food production, it is not possible to ignore the fact that there is a knowledge value to be kept and adapted to the most recent challenges related to growth. Although belonging to the low technological sectors, nevertheless, food-related productive small firms drive their business under very hard competitive conditions. This is why, in this case, as in other forms of production, the phenomenon of technological change promoted by innovation is mainly responsible for entrepreneurial success, and has been the forerunner of regional prosperity. Traditional food production represents a significant share of income for many regions, many of them rural or lagging behind within the European Union and other parts of the world. Thus, it is crucial to reinforce the acceptance of new attitudes towards modern technological paths in such areas and sectors.

References

Allen, P. and Goodman, M. (2001), 'Shifting Plates in the Agri-Food Landscape: The Tectonics of Alternative Agri-food Initiatives in California', Conference on the International Perspectives on Alternative Agro-Food Networks: Quality, Embeddedness, Bio-Politics, 12–13 October, University of California, Santa Cruz.

Avermaete, T., Viaene, J., Morgen, E.J. (2004), 'Determinants of product and process innovation in small food manufacturing firms', *Trends in Food Science and Technology*, vol. 15, pp. 474–483.

Bazoche, P., Giraud-Héraud, E. and Soler, L-G. (2005), 'Premium Private Labels, Supply Contracts, Market Segmentation, and Spot Prices', *Journal of Agricultural and Food Industrial Organization*, vol. 3, issue 1, pp. 1087–1087.

Bodin-Roudier, D. (2000), *La Guerre Alimentaire a commencé*, Paris, Albin Michel S.A.

Boyer, R. and Durand J.P. (1998), *L'après Fordisme*, Paris, La Découvert and Syros Alternatives Economiques.

Boyer, R. and Saillard, Y. (1995), 'Un précis de la regulation', in : Robert Boyer and Yves Saillard (eds), *Théorie de la régulation : L'état des savoirs*, Paris, La Découvert, pp. 58–68.

Connor, J. M. and Schiek, W.A. (1997), *Food Processing: An Industrial Powerhouse in Transition*, New York, John Wiley & Sons.

Copus, A. and Skuras, D. (2006), 'Accessibility, Innovative Milieu and the Innovative Activity of Business in EU Peripheral and Lagging Areas', in: Teresa Noronha Vaz, Eleanor Morgan and Peter Nijkamp (eds), *The New European Rurality: strategies for small firms*, London, Ashgate.

Coriat, B. and Dosi, G. (1995), 'Evolutionnisme et regulation: divergences et convergences', in: Robert Boyer and Yves Saillard (eds), *Théorie de la Régulation : L'état des savoirs*, Paris, La Découverte, pp. 500–510.

Cotterill, R.W. and Iton, C.W. (1993), 'A PIMS analysis of the Structure Profit Relationship in Food Manufacturing', in: Ronald W. Cotterill (ed), *Competitive Strategy Analysis in the Food System*, Oxford, Westview Press.

Galizzi, G. and Venturini, L. (1999), *Vertical Relationships and Coordination in the Food System*, Heidelberg, Physica-Verlag.

Guptill, A. and Wilkins, J.L. (2002), 'Buying into the food system: Trends in food retailing in the US and implications for local foods', *Agriculture and Human Values*, vol. 19, n°1, Springer, Netherlands

Hinrichs, C. (2001), 'Imputing Quality by Claiming Place: The Social and Environmental Discourse of "Local" Foods', Conference on the International Perspectives on Alternative Agro-food Networks: Quality, Embeddedness, Bio-Politics, October 12–13, University of California, Santa Cruz.

Murdoch, J., Marsden, T. and Banks, J. (2000), 'Quality, Nature and Embeddedness: Some Theoretical Considerations in the Context of the Food Sector', *Economic Geography,* Vol. 76, n° 2, pp. 107–125.

Parrot, N., Wilson, N. and Murdoch, J. (2002), 'Spatializing Quality: Regional Protection and the Alternative Geography of Food', *European Urban and Regional Studies*, Vol. 9, No. 3, pp. 241–261.

Patnaik, U. (2003), 'Global Capitalism, Deflation and Agrarian Crisis in Developing Countries', *Social Policy and Development Programme*, Paper Number 15, United Nations Research Institute for Social Development.

Sage, C. (2001), 'Entanglement, Embeddedness and the Geography of Regard: Agro-Food Networks in South West Ireland', Conference on the International Perspectives on Alternative Agro-Food Networks: Quality, Embeddedness, Bio-Politics, 12–13 October, University of California, Santa Cruz.

Scherer, M. and Ross, D. (1990), *Industrial Market Structure and Economic Performance*, Boston, Houghton Mifflin Co.

Schrader, L. and Collins, N. (1960), 'Relation of profit rates to market structure in the food industries', *Journal of Farm Economies* vol. 42, pp. 1526–1527.

Smith, A. (2006), 'Green niches in sustainable development: the case of organic food in the United Kingdom', *Environment and Planning C: Government and Policy*, vol. 24, pp. 439–458.

Soros, G. (1998), *The Crisis of Global Capitalism. Open Society Endangered,* USA, Public Affairs.

Vaz, M.T.N. (2007), 'The Design of Industrial Models', in: S. Desai, R. Stough and P. Nijkamp (eds) '*Entrepreneurship and Innovation: A Spatial Perspective*', Cheltenham, United Kingdom, Edward Elgar.

Vaz, M.T.N., Morgan, E. and Nijkamp, P. (2006), 'Rural Europe and small firms: a strategic positioning and synthesis', in: M.T.N. Vaz, E. Morgan and P. Nijkamp (eds) '*The New European Rurality: Strategies for Small Firms*', Aldershot, United Kingdom, Ashgate.

Vindigni, G., Nijkamp, P., Carrà, G. and Peri, I. (2006), 'Organizational success factors in local agro-food industries', in: Teresa Noronha Vaz, Eleanor Morgan and Peter Nijkamp (eds), *The New European Rurality: strategies for small firms*, London, Ashgate.

Weiss, L.W. (1974), 'The concentration-profits relationship and antitrust', in: H.J. Goldschmid, H.M. Mann and J.F. Weston (eds) *Industrial Concentration: The New Learning,* Boston, Little Brown.

Chapter 17
Traditional Food Production, Market Segments and Rural Sustainable Development: A Synthesis

Jean-Louis Rastoin

The question implicit in the title of this work is eminently political and strategic: political because it refers to the governmental decisions concerning the orientations and incentives observed in the food system, both in terms of the consumption model and the supply model, taking into account the three pillars of sustainable development (ecology, economy, equity); and strategic because, for the parties involved in the food systems, it will lead to choices which differ significantly according to their size and position in the food commodity chains.

Indeed, numerous countries throughout the world which adhere to the Rio charter on the environment or to the *Millennium Ecosystem Assessment*, or which signed the Kyoto Protocol, and even some of the major non-signatories (for example, California in the United States) have – to a greater or lesser degree – already introduced an 'environmental protection' section into their policies and are making progress with regard to 'equity' or 'social responsibility'. It goes without saying that all countries around the globe are convinced that, without direct or indirect economic viability (by means of inter-category transfers) no market activity can be envisaged.

We are witnessing, therefore, the gradual emergence of a legislative and regulatory framework aimed, on the one hand, at protecting our natural heritage and reducing the most extreme social asymmetries and, on the other, at maintaining or increasing competitive intensity across sectors. It is clear that three parameters of sustainable development are interdependent, and that a balance must be found between them. However, a new element to be considered is the doubtless irreversible emergence and consolidation of the first two parameters, and this should assume an ever greater importance in the strategic decisions of all actors, be they producers or consumers. For example, with regard to firms, the appearance of independent agencies which assess and rate the two of three aspects of sustainable development (ecology and equity)[1] is leading to more and more radical changes to managerial practices. The rating, which is implemented on the basis of increasingly sophisticated indicators, is either published, whereby

1 Economic evaluation has long been practised by financial auditors.

the firms are left to the judgement of public or professional opinion, or requested by certain partners (for example, investment funds), or it may even become a communication tool for firms. Despite the sophistication of the messages communicated by firms, ensuring that their words remain far from reality, and the sometimes abusive nature of certain messages, we can nevertheless assume that these 'signals' conveyed by the ratings imply a slow but sure change of direction towards sustainable development.

This general context applies to all sectors. But what is the situation in the food system?

First of all, we must remember that this system is very heterogeneous and, in all countries, assumes at the very least a dual nature, with an agro-industrial subsystem concentrated, specialized, financialized and integrated in the globalization movement based on a rationale of international competitiveness; and a more traditional subsystem based on smaller territories and thus smaller firms with a rationale relating more to proximity.

Under the constraint of sustainable development, the agro-industry model endeavours to adapt in the following ways: by emphasizing economies of scale (size effect), thereby allowing prices to be kept low and solutions to be found in terms of environmental protection (structured agriculture, controlled industrial processes); by renewing its product portfolio through the provision of food hygiene; and by improving the logistical provisions in order to maintain the competitiveness of very long commodity chains (food miles). Massive investment in both material and immaterial (R&D, communication, know-how) infrastructures is possible because of the size of the firms. This model satisfies the expectations of most consumers thanks to the model of society which has tended to dominate throughout the world since the fall of the Berlin wall in 1989. However, there is no guarantee that it is capable of taking up the double challenge of this millennium. The first challenge is that of public health, i.e. providing non-pathogenic food to 9 billion human beings in 2050. It has in no way been established that producing artificial foods is a solution to combating illness originating from food or, more importantly, that this solution can be applied on a worldwide scale. The second challenge concerns resources. The generalization of a consumption model rich in meats seems problematic because of the limited potential of farmland. Moreover, because of its intensiveness in terms of inputs (in particular water and chemical products) and energy, the agro-industry model is restricted by the physical limits of the availability of water and oil.

The second subsystem, which we describe as 'alternative', is founded on very different – and sometimes opposite – bases than those we have just considered: short commodity chains from a production point of view (transformation of local raw materials) and the use of traditional know-how bestowing a *typicité* and high organoleptic qualities on the products; a strong cultural element to products; and the social involvement of consumers in the context of a diversified food model. The economic viability of the proximity model implies that customers are willing to pay more for food. It requires prior knowledge of the products and a greater time investment when planning and preparing meals. Finally, this model requires the respect of health standards during production and a high degree of marketing skill, demonstrating that it is not simply a question of a nostalgic return to ancestral practices. The alternative food system, like the agro-food system, must respect those constraints of public health and sustainable development. This means mobilizing the most up-to-date scientific and technological resources.

This book suggests several trains of thought which support the hypothesis that the alternative model has prospects for development.

First, sustainable development concerns are analysed in detail in the meta-analysis of the agro-environmental policy of the European Union (Oltmer et al.), with a very strong conclusion: following significant results with regard to the reduction of pollution by nitrates, the biggest challenge is still to come with regard to the reduction of the animal density per hectare and the extension of the grasslands, both of which are essential to alleviating the pressure of agricultural activity on the natural environment.

Further diagnosis is supplied by the comparative analysis of rurality at European level, even comparing it with Turkey (Gülümser et al.) and by the description an evaluation of the governance structures and the determinants of local economic development in Raia Ibérica (Neto et al.).

The cultural values linked to diet and its role in rural development is clearly highlighted by Hungarian 'gourmet festivals' which combine popular arts, in particular music, with food festivals (Szlanyinka). The interest of this type of event is to perpetuate the collective memory and to attract tourists from outside the local community looking for authentic experiences. It is nevertheless important to remain cautious concerning the 'modernization' of such events: 'Disneyfication' in 'Gastroland' would run against the expectations of potential customers. The necessary progress in organizing and running gourmet festivals must not lead to the loss of cultural identity.

Five other products studied in this book confirm the potential of food products anchored in tradition and which are differentiated by means of quality, thereby enabling them to withstand competition from agro-industry products: meat in Latin America, bread in Romania, honey in Europe, olive oil in Mediterranean countries, and beer in Belgium.

Meat-based products in Latin America present a very interesting case of cultural and technological exchange. Introduced during the Iberian conquest in the 16th century, they underwent a process of cultural and technical adaptation enabling a wide variety of local specialities to be maintained and thereby to withstand the subsequent competition from standard North-American products. Today, they represent an opportunity in the form of a reservoir of innovation and ethnic products for expatriates (Mateo et al.)

In Romania, traditional bakeries face competition from bread factories and suffer from the concentration of flour mills, which exercise a certain market power (Ionel). However, the quality differential and the social environment of the products means those small bakeries can withstand this competition, as shown by the cases of France or Italy.

Honey is a complex hybrid product which requires sophisticated technologies deployed in micro-firms. It is often a secondary activity providing farms with additional cash flow. The differentiation of the types of honey produced, resulting from the plants from which pollen is gathered and hence from the area of production, can be used to its advantage. From an environmental point of view, it is an indicator of plant pathologies and the bees are indispensable intermediaries in pollination. Finally, it is also of social interest, as it often stimulates associative forms for production and commercialization (Noronha Vaz).

Olive oil is a product typical of both an agro-climatic zone and a civilization in the Mediterranean basin. The Mediterranean area currently has almost a monopoly, and international demand is growing rapidly because of the success of the traditional Cretan diet. The olive tree is particularly well adapted to the dry conditions and poor-quality soil which characterize huge areas of the Mediterranean countries (Mili). Mediterranean olive oil is nevertheless under threat from several factors: high production costs; sometimes insufficient quality; the market power of supermarkets; gaps in the information systems; and, in particular, large plantations in the countries of the new World (Australia, Chile, and California) whose efficiency in terms of marketing should not be underestimated (cf. the wine sector). Consequently, an alliance strategy of Mediterranean producers focusing on the creation of a regional label and the creation of technical and commercial networks would appear to be necessary.

In Belgium, we witness a confrontation between traditional beers and beers marketed by large firms (Avermaete and Vandermosten). The former are threatened by the evolution of consumers' tastes, compounded by huge communication budgets and food safety constraints. A strategy based on product, process and organizational innovation, as well as a labelling policy enables these microbreweries to withstand the competition and even to increase sales.

Consumption tendencies in the European context have also been analysed: the Romanian food consumption model in the context of the European Union integration (Alexandri and Alboiu); the growing market power for butter (Hockmann and Vőneki); and the study of the bio-products market in eight European countries (Thøgerson). This last study demonstrates the homogeneity of the determinants of consumption of this type of product, with three vectors: the health benefits; taste quality; and environmental protection. Furthermore, the survey reveals that there is no rejection of transformed bio-products. These elements support both the alternative production model (protection of natural heritage) and the consumption model (diversity of products including prepared foodstuffs, taste).

The growth potential of this type of products appears to be substantial, whereas agro-industrial products are currently demonstrating a global tendency towards stagnation. On the basis of statistics available at a European level for 'labelled' food products and in particular for products with an 'identification of geographic origin' (IGO), we can assume that these products will continue to demonstrate considerable rates of increase on the food market in the coming years.

Amongst the factors conditioning these growth prospects for products carrying an IGO, we must not omit the relationship with the respect for the environment, as shown in Chapter 10 in the example of the DOC (guaranteed quality labels) in Emilia-Romagna (Gatti). Furthermore, this relationship is of great importance in the international negotiations (WTO) as it enables tensions concerning the use of geographic labels as a commercial brand to be overcome. The DOC guarantees the traceability of the product. We then return to the concerns of sustainable development.

All these examples show that there are both threats and opportunities in the alternative food commodity chains of the agro-industrial model. In addition to what the study of previous cases has shown (technical model adapted to the regional context, consumer's interests), we should also consider two important aspects, i.e. anchoring activities in rural areas, even though these areas are becoming increasingly deserted, and supporting the fight against the deterioration of the natural environment. Finally, local development could be systematically encouraged because, thanks to the regionalized food commodity chains, it is possible to boost tourism through the environmental, cultural and gastronomic heritage of the regions, thereby facilitating their insertion at the international level. The key factor of success of such a strategy is to respect the regional identity which is the cornerstone of differentiation in a world tending towards standardization.

However, the forces at work are unequal, and it is impossible to envisage the preservation and subsequent expansion of the alternative model without the existence of a political will. Global standardization of food consumption is gradually eliminating the regional models (thus the Cretan model disappeared 20 years ago). It is therefore urgent that we implement a programme to establish food memory at the regional level, as well as an ambitious education programme starting at primary school level. This means making young children aware of the culinary heritage of the region in which they live, with a view to encouraging dietary and social behaviour which differ from that resulting from the mass consumption model. This type of action should be accompanied by incentive measures promoting collective chains. The public authorities must also organize the transfer of technological and managerial know-how to the firms, as well as support the development of both horizontal (creation of baskets of complementary products) and vertical (pooling resources) networks. Incentive measures can involve financial, fiscal or technical support and must be subject to the dual conditionality of traceability (use of local raw material) and environmentalism (non-damaging production methods).

If we consider a geopolitical context of a multi-polar world in one generation's time (2030), a probable scenario includes three poles, one of which is already established (the United States), the second is emerging (China), and the third is still somewhat vague (the European Union).

Simplifying this, we could suggest that the United States could draw its economic power from high technology and services, while China would draw its power from the industry of mass consumption goods. For Europe, uncertainties remain. We might postulate that the food system, with its considerable diversity, its high level of both qualitative and symbolic differentiation, the depth of its historic roots and the specificity of its institutional context of territorial foundations, is one of the cornerstones of the model of development.

Index